ULTRA-METABOLISM

THE SIMPLE PLAN
FOR AUTOMATIC WEIGHT LOSS

MARK HYMAN, MD

COAUTHOR OF THE *NEW YORK TIMES* BESTSELLER
ULTRAPREVENTION

© 2008, 2006 by Mark Hyman, MD

Portions of this book were previously published in 2006 by Simon & Schuster.
The Rodale Inc. direct mail edition is published in 2008
under license from Simon & Schuster.

Library of Congress Cataloging-in-Publication Data

Mark Hyman.
    UltraMetabolism : the simple plan for automatic weight loss / Mark Hyman.
        p.   cm.
    "Portions of this book were previously published in 2006 by Simon & Schuster."
    Includes bibliographical references and index.
    ISBN-13 978–1–59486–654–8 hardcover
    ISBN-10 1–59486–654–6 hardcover
    1. Weight loss.   2. Genomics.   3. Metabolism.   I. Title.
RM222.2.H964 2008
613.2'5—dc22   2008024841

2 4 6 8 10 9 7 5 3 1 hardcover

We inspire and enable people to improve their lives and the world around them
For more of our products visit **rodalestore.com** or call 800-848-4735

For
all those whose bodies came without an instruction manual

For
my patients who taught me to ask the questions
that allowed me to discover their unique instructions

For
my wife, Pier, my children, my parents, and my family,
without whom my life would be very empty

CONTENTS

Many obese people would like to reduce their weight or would benefit considerably if they were able to do so. But they will not be helped by relentless moralizing and easy solutions reflecting a theory of gluttony that does not stand up to the available evidence.

—WILLIAM IRA BENNETT, M.D.[1]

With obesity, we are thus generally not facing problems with gluttony, poor willpower or laziness, which is unfortunately a prevailing view even among highly educated, specialized physicians. Instead, a very slow, limited, continuous positive energy balance is responsible. Unfortunately, the construction of the ancient human organism is such that it favors this development by powerful, normal mechanisms developed during evolution.

—PER BJORNTORP, M.D., PH.D.[2]

I am angry and frustrated.

We are in the middle of a medical revolution that has finally unearthed the keys to permanent weight loss. This dramatic breakthrough can help fix the rampant obesity problem that's affecting millions of Americans today.

But I am angry because virtually no one knows about or is acting on the powerful information revealed by this revolution.

I am frustrated that the medical profession, our government, and the food industry have ignored this information, either on purpose or through negligence.

I know because I have used this cutting-edge, yet hidden knowledge for nearly ten years at one of the world's top health resorts, Canyon Ranch, and in my private practice to help thousands of people lose weight and keep it off for good.

This position has allowed me to perform innovative testing to peek into the inner workings of the body in order to discover the keys to health and weight loss. I've spent years analyzing the wealth of data from these powerful tests. I've had a unique opportunity to try different methods with my patients to discover what works and what doesn't work for permanent weight loss.

And I have been fortunate enough to connect the weight loss dots in a way that, to my knowledge, has never been done before. UltraMetabolism makes that knowledge available to everyone for the first time.

This medical revolution is based on *nutrigenomics*—the science of how food talks to your genes—and promises to turn up your metabolism, help you lose weight, keep it off, and get healthy for life.

The reason many people continue to struggle with weight loss is that their recommendations have not been personalized. There is no one perfect diet, supplement, pill, or exercise program for everyone. That is the exciting promise of the latest scientific research on our genes, nutrigenomics.

In this book I shatter the confusing and misleading myths that make you gain weight and keep it on. You'll be surprised to discover that much of what you know about dieting and weight loss simply isn't true and has, in fact, actually been making you gain weight.

I give you the keys to use the information and knowledge at the heart of this medical revolution in the form of a simple plan that will help you lose 11 to 21 pounds in the first eight weeks of the program.

Many of my patients have experienced much more dramatic weight loss than this, some less, but on average, that's what you can expect.

In part I, I explore all the myths that confuse, confound, and mislead us in our struggle to lose weight and get healthy.

In part II, I give you the keys to turn on your metabolism, turn on your fat-burning genes, turn off your weight gain genes, and program your body to lose weight automatically.

In part III, I detail a simple eight-week plan to help you lose weight based on your unique genetic needs. Since each of our bodies is different and may require more or less of certain nutrients to awaken our fat-burning DNA, I show you exactly how to customize the program for your own particular needs.

The program includes menus, recipes, and shopping lists, as well as recommendations for supplements, exercise, and lifestyle treatments designed to create a healthy metabolism and lifelong health.

The secret medical revolution explains why we are experiencing an unprecedented epidemic of obesity and chronic health problems. It provides a new road map to navigate our way back to health, fitness, and successful long-term weight loss without starving ourselves or counting calories, fat grams, or carbs. We simply have to eat in harmony with our genes.

INTRODUCTION

We Are Designed to Gain Weight

The human body is designed to gain weight and keep it on at all costs. Our survival depends on it. Until we face that scientific fact, we will never succeed in achieving and maintaining a healthy weight. Doctors and consumers alike believe that overeating and gluttony are the causes of our obesity epidemic. Science tells a different story: it is not your fault you are overweight.

Powerful genetic forces control our survival behavior. They are at the root of our weight problems. Our bodies are designed to produce dozens of molecules that make us eat more and gain weight whenever we have the chance.

Furthermore, the food industry and our government's recommendations are fueling this feeding frenzy. We cannot expect to change our instinctual responses to food any more than we can eliminate a feeling of terror when confronted with danger.

We have evolved over hundreds of thousands of generations under conditions of food scarcity. The genes and molecules that control our eating behavior were shaped by those times. Our DNA was designed for accumulating fat in the days when we had to forage for food in the wild.

The body's weight control systems were designed to make us gain weight, not lose it. Our bodies and our DNA evolved in an environment of food scarcity, not overabundance. Ignoring that fact is hazardous to both our health and our waistlines.

Genes and Obesity: The Power of Nutrigenomics

This book will help you understand what science has to say about the forces that guide your eating behavior and weight control. It will give you the tools to make that knowledge work with your genes, not against them. It is the new science of weight loss, based on the new field of nutrigenomics. This is the science of how food and nutrients interact with our genes to turn on messages of health or disease, of weight gain or weight loss. It is not about finding the right diet.

There is no one solution for everyone, a one-size-fits-all weight loss strategy that works all the time for every person.[1] It is about finding the right diet for you, based on understanding the unique ways in which your genes and metabolism interact. The 7 Keys to UltraMetabolism make this revolutionary field of nutrigenomics, of personalized medicine and nutritional healing, accessible for the first time.[2]

Despite the $50 billion spent on weight loss every year in America, there is no magic diet, pill, or exercise routine that is guaranteed to make you lose weight. Most diets fail, and Americans grow larger and more frustrated by the minute. Obesity⋆ rates have tripled since the 1960s. More than two thirds of the adult population and one third of our children are now overweight.

Obesity is overtaking smoking as the number one cause of preventable deaths in this country, accounting for increasing rates of heart disease, stroke, cancer, dementia, and diabetes.[3] If you are overweight at 20 years of age, your life expectancy is 13 years less than that of your contemporaries of normal weight.[4]

Only 2 percent to 6 percent of all attempts at weight loss are successful.[5] And the average person ultimately gains 5 pounds over and above his/her starting weight for every diet he or she goes on. Dieting does not work.

A recent comparison of different diets (Atkins, Ornish, Weight Watchers, and the Zone) published in the *Journal of the American Medical Association*[6] found no significant differences between them and an average of only 2 to 3 kilograms of weight loss after one year. In the accompanying editorial, Dr. Eckel concludes, "It seems plausible that for maintenance of reduced body mass, the right diet needs to be matched with the right patient. Ultimately, a nutrigenomic approach most likely will be helpful."[7] Our DNA holds the secrets to getting us off the endless wheel of diets and inevitable failures.

How many of you have lost and gained weight over and over in a frustrating and discouraging cycle? We can succeed only by working with our genes rather than against them to create a desirable weight and stay there. We must stop falling for the myths that keep us fat and perpetuate an endless cycle of weight loss and gain. Using the 7 Keys to UltraMetabolism, you can learn how to awaken your fat-burning DNA.

What Is Metabolism?

What is your metabolism, and what does it have to do with your weight? Most people think of metabolism as the part of your biology that controls how fast or slow you burn calories. We say, "She has a slow metabolism" or

---

⋆ Overweight is defined as a body mass index, or BMI, of >25 (BMI = weight in kilograms divided by height in meters squared). Obesity is defined as a BMI of >30. Go to www.ultrametabolism.com to determine your BMI.

"He has a fast metabolism" when accounting for why someone is over- or underweight.

In general, that is true; however, metabolism is actually the total of all the chemical reactions in your body, the finely choreographed dance of molecules that determines health or illness. Metabolism, as I refer to it in this book, is meant to describe all the molecules, hormones, and brain, gut, and fat cell messenger chemicals that regulate, among other things, our weight and the rate at which we burn calories.

Our eating behaviors, the quality of our food, and changes in our environment, stress levels, and physical activity influence the way in which our metabolism processes food, burns calories, and regulates our weight. The 7 Keys to UltraMetabolism describe the various parts of our metabolism that are affected by our environment, our genes, and our habits.

Scientific advances over the last few decades have allowed us, for the first time, to understand how our bodies work, to understand how diet really influences our genes. That knowledge is the key to creating a healthy metabolism.

Your Body's Owner's Manual

This is not a diet book. This is your body's owner's manual. It contains the instructions on how to regulate the basic functions of appetite and metabolism. Understanding what controls our appetite and metabolism (the way our bodies use or burn calories) is the key to success. Once we learn them and begin to fit into our *genes,* we will finally and forever fit into our *jeans.*

My Own Story

Even though my weight has never fluctuated by more than 10 pounds, I have struggled with food and watched its effects on me; the stress and struggles of medical training, working in the emergency room, and chronic illness pushed me to my limits. I found my energy and mood swing wildly as I downed a triple espresso, a giant chocolate chip cookie (or two), and half a pint of Ben & Jerry's ice cream just to keep myself going before each emergency room shift.

Struggling with my own health, food, mood, and energy issues and suffering and recovering from chronic fatigue syndrome have given me more compassion for those who struggle with the chronic illness of obesity or with their weight.

To help my patients and myself, I have read thousands of scientific papers to understand how the body actually works. In my years as co–medical director at Canyon Ranch and now in my medical practice, I have also treated thousands of patients and helped them lose thousands of pounds.

I am often the last stop on the road for my patients after they have tried every popular weight loss diet, pill, and surgery and experiencing repeated failures. In order to help them, I am compelled to determine why they failed. They also often come to me with a "whole list" of medical problems in addition to being overweight. That is why I refer to myself as a "whole-listic" doctor.

My recovery from another chronic illness—chronic fatigue syndrome, a story I told in *UltraPrevention*—has given me more insight into dealing with chronic health problems. After running myself into the ground working in the emergency room, going through a traumatic divorce, absorbing toxic levels of mercury during a year in Beijing, China, and finally suffering from food poisoning, I became severely ill.

Struggling with and fully recovering from a chronic problem that is often considered incurable (as obesity is), I learned how to work with every system of the body—my own body. And last, I have experimented with many different ways of eating: lots of sugar and coffee, junk-food vegetarian, whole-food vegetarian, high-protein, low-carb, high-fat, and everything in between, and to a lesser degree than those with a strong genetic predisposition, I have observed the power of food to affect weight, energy, and mood.

Food is a drug. Food is medicine. Hippocrates taught us this centuries ago. Whether it is good or bad medicine depends on how you use it.

UltraMetabolism is the result of all my experiences and knowledge—with my own struggles with food, energy, and illness, with my patients and my academic learning. It is not a magic cure or a quick-fix weight loss diet or fad. It is a way of living. It is a way of living in harmony with your genes. It is the next step in the evolution of our knowledge about how to care for ourselves and the first approach that translates recent scientific advances into a practical program.

The Secret Twenty-first-Century Medical Revolution

A new scientific truth does not triumph by convincing its opponents and making them see the light, but rather because its opponents eventually die, and a new generation grows up that is familiar with it.

—MAX PLANCK (1858–1947)[8]
Nobel Prize winner, quantum physicist

The revolution sweeping medicine today is rooted in our DNA, our unique individual genetic structure. Most patients and many doctors are as yet unfamiliar with the genomic revolution and how it applies to our weight. The genomic revolution opens a new window into why our bodies do what they do and how we can work with our genes rather than against them.

The mapping of our DNA allows us to understand the tremendous variation of genes from individual to individual. The Human Genome Project has identified the 8 billion letters that make up our 30,000 genes. But the meaning of the letters is still being deciphered, like an ancient scroll or hieroglyphic.

The human species has approximately 3 million variations in those letters, which is what really makes each of us unique. Scientific understanding of how those genes and their differences from person to person influence metabolism is deepening daily. The future of medicine lies in cracking that code—the way in which genes *(genomics)* create proteins *(proteomics)* and proteins control our metabolism *(metabolomics)*. Translating that science into practical strategies for health, longevity, and weight loss is the promise of UltraMetabolism.

The interaction among our genes, our diet, and our environment provides clues to the hidden mystery of why our bodies malfunction and how to help them achieve optimal functioning. Recent scientific advances provide deeper insight into the inner workings of our bodies. For the first time in history, we are able to understand the owner's manual for each and every one of us.

Translating Scientific Research into Medical Practice

Despite significant advances in biology over the past two decades, much of this knowledge has yet to be translated into medical practice.

—SATYANARAYAN AND SHOSKES[9]

Researchers and clinicians have yet to apply this new knowledge to helping people achieve and maintain a healthy weight. While the pieces of this puzzle are increasingly being discovered, they have yet to be effectively synthesized and made practical.

Scattered in numerous laboratories and research papers are the latent answers to the most profound health problems in this century. Dr. Bill Frist, the U.S. Senate majority leader, recently lamented, "It takes our physicians an average of 17 years to adopt widely the findings of basic research."[10] The

information in this book is not at work in doctors' offices, public policy
weight loss programs, or other diet books. Practice has not caught up with
science. Information has not become practical knowledge. It has not been
turned into a prescription for weight loss or health. That is why I wrote this
book—to make this revolution available to you today.

The UltraMetabolism Prescription: Turning on the Fat-Burning Genes

The UltraMetabolism Prescription helps you create health and lose weight
by putting these fragments of information into a comprehensive explanation and practical strategy. These principles, developed during my twenty
years of medical practice, represent a fusion of traditional medical techniques with practical approaches learned from working with my
patients.

In the first ten years, I relied more strictly on techniques learned in medical school to help my patients lose weight, maintain normal blood pressure,
control blood sugar, and feel better. In the last ten years, I have learned more
directly from my patients about what works and what doesn't.

I have had the privilege of listening to thousands of stories, connecting
them with thousands of tests exploring every aspect of their metabolism,
and making new connections and discoveries that have tied together the
science, the patients, and their results. During this time, I have helped thousands of patients lose thousands of pounds in my work at Canyon Ranch,
one of the world's finest health resorts and wellness centers, and in my private medical practice.

Many of the answers to our weight and health problems are buried in
thousands of research papers. Most practicing doctors have little time to research and synthesize these findings. But I have read them, digested them,
seen the bigger picture of how our genes interact with our diets, and put
these findings into practice.

I have, with the aid of a number of brilliant thinkers and scientists, synthesized and researched the greatest advances in medical science over the
last twenty years, advances that usually take decades to be incorporated into
medical practice. Unfortunately, these breakthroughs are not available to
the average person. These breakthrough concepts and understanding of the
basic laws of biology, the laws of nature, and how they affect our weight and
metabolism are the basis of UltraMetabolism.

Through revolutionary new types of testing and working with the
world's best nutritionists, exercise specialists, and behavioral and stress management experts, I have repeatedly witnessed how the concepts outlined in

this book can change people's lives forever. It is analogous to giving them the key to their dream house—a trim body that works well, stays fit, and feels fabulous.

In my last book, *UltraPrevention,* which I coauthored, I outlined the model for a novel way of thinking about health, one based on the underlying causes of disease. As we know too well, conventional medicine focuses on the symptoms of a disease, not its cause.

In the case of obesity, the most effective drug we have is Xenical, which blocks fat from being absorbed and gives you horrible stomach pain and uncontrollable diarrhea if you eat fat. It also causes significant hormonal and nutritional imbalances that result in more rapid aging. A new drug, rimonabant, blocks the cannabinoid receptors (special receptors in the brain that when activated cause the munchies as if you were smoking pot), but its side effects can include depression, anxiety, nausea, and uncontrollable munchies when you stop taking the drug.

There is no magic pill. In order for any weight loss drug to be effective, it must be taken forever. This will not necessarily correct all of your metabolic problems or improve your health. It is simply a woefully misguided approach to losing weight.

The right approach is to find and treat the underlying causes of obesity and weight gain. My secret is that I never tell my patients to lose weight. I help them understand their bodies and how to work with them rather than against them. I help them discover the factors that impair their health and to remove them, then help them discover what they need to thrive. The weight loss happens effortlessly and automatically. That is the promise of UltraMetabolism.

PART I

The 7 Myths That Make You Gain Weight

⊹

There are seven myths that many of us believe that confuse, confound, and thwart our efforts to lose weight. These beliefs are impediments to success. One by one I will shatter these myths that make you gain weight and keep it on.

1. The Starvation Myth: Eat less + exercise more = weight loss
2. The Calorie Myth: All calories are created equal
3. The Fat Myth: Eating fat makes you fat
4. The Carb Myth: Eating low carb or no carb will make you thin
5. The Sumo Wrestler Myth: Skipping meals helps you lose weight
6. The French Paradox Myth: The French are thin because they drink wine and eat butter
7. The Protector Myth: Government food policies and food industry regulations protect our health

We cause trouble for ourselves by believing these myths. Understanding how they contribute to weight gain will free you from an endless series of habits and beliefs that prevent success in achieving your ideal weight and optimal health. The lessons learned from exploring these myths will give you practical tools to change the habits and behaviors that undermine your weight loss and health goals.

CHAPTER 1

THE STARVATION MYTH:

Eat Less + Exercise More = Weight Loss

"I Don't Eat Much, and Still Can't Lose Weight"

*J*oanna, *a fifty-three-year-old executive secretary and mother of three, was a bundle of frustration and resignation. She recounted a story I had heard from hundreds of patients. She had tried every diet; liquid diets, low-carb diets, low-fat diets, "fad" diets, shakes, prepared meals, and raw foods. And every time the same thing had occurred: she had lost weight initially, only to gain it all back, plus some more. Now, she said, she didn't even eat that much and she still couldn't lose weight.*

My initial training would have led me to believe she was either deceiving herself or hiding her eating habits from me, but after many years of blaming the patient for lack of willpower, for overeating and not exercising enough, I realized that there must be more to the story.

By the time Joanna came to see me, I knew better and had her complete a detailed three-day diet record—everything she put in her mouth for three days. In fact, she was constantly dieting on poor-quality food, putting her body into a state of chronic low-grade starvation.

Then I tested her resting metabolic rate, or RMR, the amount of calories the body burns at rest. Not surprisingly, I found she had a slow metabolism—she burned fewer calories than expected for her age, sex, height, and weight. Further testing of her body composition—what percentage of her body was made up of fat—showed that she was underlean (this often goes along with being overfat, but not always). In other words, she had less muscle than expected for her body weight.

Why? Her repeated restricted-calorie or partial starvation diets had led to loss of both fat and muscle. When she had gained the weight back—as she always did—she gained back all fat, leading to a slower metabolism. Fat burns 70 times fewer calories than muscle. So she needed a lot fewer calories to sustain her increased body weight. By following the eat less = weight loss model, she had actually made things worse for herself.

Getting her to eat the right foods—whole, unprocessed ones—and in quantities that exceeded her resting metabolic rate led to steady and sustained weight loss, get-

ting her off the yo-yo, seesaw, merry-go-round of diets. As I do with many of my pa-
tients, I advised her, paradoxically, to eat more, *not* less, *to lose weight.*

It's Not Your Fault You Are Overweight

We've all heard the fiendishly simple and completely untrue "truths" that abound in our culture that "teach" people what they need to do to lose weight: "Eat less and exercise more," "Stop eating so much," "It's all about willpower," "Everyone knows that people who are overweight are lazy, undisciplined, and self-indulgent." Most of you who are trying to lose weight have internalized a cultural message that it's *your* fault you're fat.

When I first started my medical practice, I believed the formula for weight loss was simple: Eat less + exercise more = weight loss. I thought the only reasons people couldn't lose weight were that:

1. They overate.
2. They were lazy and didn't exercise enough.
3. They ate too much *and* didn't exercise enough.

Now I know better. For many overweight and obese individuals, these explanations are overly simplistic. After almost two decades in practice, I know that relying on this myth of weight loss is terribly inadequate. Not only is it completely unsupported by the scientific literature, it creates a blame-the-victim mentality that tells people who are struggling with their weight in a not-so-subtle way that if they only tried harder they would lose weight. There is only one problem with this point of view: it's not true.

Nobody Wants to Be Fat

The truth is that nobody wants to be fat. It isn't your fault that you have problems with weight. It isn't as though you chose this struggle. Even if you had enough willpower to keep yourself from eating when your body tells you to, you might still not be able to lose weight. This is because our bodies are genetically wired to make us gain weight and keep it on. You cannot get away from this basic biological, evolutionary truth, but you can do something about your weight.

The trick is learning how to tune up your metabolism and use your body's natural calorie-burning capabilities to help you lose weight and get healthy. But believing the common oversimplifications about your weight that our culture promotes won't help you to sustain long-term weight loss and health. You need to understand that the human body is much more complex than this.

What makes us thin, fat, or somewhere in between does have something to do with how much we eat and exercise. But the oversimplification stops there. Complex forces that govern our survival control our weight and metabolism. In fact, there is no one simple reason why an individual may have trouble with his or her weight.

Over the last ten years, medical research has revealed that weight loss is much more sophisticated than our outdated preconceived notions about eating less and exercising more. We have found there are seven basic principles involved in metabolism that affect weight regulation to one degree or another.

UltraMetabolism is founded on these seven principles, and in part II of this book you will learn about each of these principles, or keys to weight loss, as well as practical ways to implement changes in each area so that you can optimize your health and lose weight. This new science of metabolism and weight loss carries us far beyond the old idea that losing weight is simply a matter of eating less and exercising more.

This analogy makes the obvious even clearer: telling someone who's overweight that all he or she has to do is eat less and exercise more is like telling someone who's poor that all he or she has to do is make more and spend less. These kinds of simple equations ignore many factors.

One of the factors left out in such simple equations is what I call the starvation syndrome. When you buy into the idea that eating less will make you lose weight, you convince yourself that you need to restrict calories from your diet. In fact, most of the popular diets on the market recommend that you do just that. The problem is that calorie restriction almost always backfires. The reason? Your body thinks it's starving to death and sets off chemical processes inside you that force you to eat more. This, in essence, is how the starvation syndrome works. Let's look at it in more detail.

The Starvation Syndrome: The Real Problem with the Old Way of Looking at Weight Loss

The World Health Organization (WHO) classifies a diet containing less than 2,100 calories a day for the average man and 1,800 calories for the average woman as a starvation diet.

The average woman dieting in America is trying to eat less than 1,500 calories a day. That means she is constantly in a state of starvation. Our culture praises starvation and excessive weight loss. We live with the starvation syndrome every day.

Fashion models today are 25 percent thinner than models forty years ago. To achieve this, many binge, purge, use laxatives, smoke, drink diet soda, and overexercise. These activities trigger a cascade of molecules in the blood designed to make them rebound from their severe diets and overeat. Thus they binge, purge, and diet even more rigorously. This turns into a very ugly cycle that works in opposition to their body's natural chemical construction.

This molecular cascade is the body's way of saving itself from starvation. Human beings are genetically coded to do this. The most fundamental parts of who we are as biological creatures are designed to keep us from starving ourselves. It is a very basic survival mechanism. These models are receiving commands from their bodies to eat more and gain back the weight they have lost for their own good. Their bodies think they are in danger, and they are sending them signals to eat so they can save themselves. Ignoring these signals makes their bodies grow old before their time. This is not a good pattern.

You may have fallen into this trap yourself. If you have tried dieting and restricted your eating below the number of calories you need to make your body function properly, you have set off the same molecular cascade inside yourself. As a consequence, you receive hunger signals that are too strong to ignore. You rebound and gain back the weight you have lost. In most cases, you gain back *more* than you initially lost. You end up on the classical weight yo-yo. Welcome to the starvation syndrome.

Your Lizard Brain: The Reason You Are Wired to Get Fat

But this all seems backward, doesn't it? If it is true that we are genetically engineered to gain weight, then it would seem that we are wired incorrectly. Why would we be designed to overeat and grow fat?

It all comes down to the oldest and most primitive part of our brain, our limbic, or "lizard," brain. This is the part of your brain that evolved first, and it's like a reptile's brain. It governs your survival behaviors, creating certain chemical responses that you have no conscious control over.

There are three basic survival behaviors controlled by our primitive brains. They are (1) our fight-or-flight response, (2) our feeding behavior, and (3) our reproductive behavior.

The first is the *fight-or-flight response.* It is a set of chemical, physical, and psychological responses that allows us to cope with dangerous or life-threatening circumstances. In humans' early history, this response was developed so that we could escape attacks from ravenous wild beasts.

I recently experienced this firsthand while on a walking safari in Africa, tracking a black rhinoceros. They are nearly extinct and may not be shot if they charge—and they frequently charge. Our guide was experienced, found the rhino easily, and then it charged us. My heart began to race, and my breath quickened. I felt the blood rush through my veins with a surge of anxiety and fear that fortified me with the strength of a lion and the ability to run like the wind. Our guide shouted and waved his arms wildly. Thankfully, at the last moment the rhino spun away.

During that experience, I had absolutely no conscious control over the chemical reactions that governed my fight-or-flight response. Why? They were hardwired survival responses. I was in danger, and my body was doing what it was programmed to do—save my life.

The same thing happens to us regarding food. The same part of your brain that controls the fight-or-flight response controls your feeding behavior. While you might think you are in complete control of your mind, the truth is that you have very little control over the unconscious choices you make when you are surrounded by food.

The key to a healthy metabolism is learning what those responses are, how they are triggered, and how you can stop them. You don't want to put yourself in the position of resisting the lure of a bagel. Your drive to eat it will overwhelm any willpower you might have about losing weight. It is a life-or-death experience, and the bagel will always win.

So one of the most important principles of weight loss is never to starve yourself. The question is whether or not you are eating *enough* calories, not whether or not you are eating *too many*. What you need is a baseline for how much you have to eat to keep your body from going into starvation mode.

The Reason Most Diets Fail

The reason diets backfire almost all the time is because people restrict too much. That is to say, they allow the number of calories they consume to drop below their resting metabolic rate. This is the basic amount of energy or calories needed to run your metabolism for the day.

For the average person it is about ten times your weight in pounds (I weigh 180 pounds, so my resting metabolic rate is 1,800 calories). This is the bottom line for your body every day if you don't get out of bed or expend any energy. (See www.ultrametabolism.com for an easy guide to calculating your resting metabolic rate.)

If you eat less than that amount, your body will instantly perceive danger and turn on the alarm system that protects you from starvation and slows your metabolism. As a consequence, you go right into starvation mode and just start eating and eating once you inevitably stop the diet—the classic rebound weight gain.

Just think of what happens when you skip breakfast, work through lunch, and finally return home in the evening: you eat everything in sight. Then you feel stuffed, sick, and guilty and regret ever entering the kitchen in the first place.

Why would you possibly want to overeat and make yourself sick? Most of us are reasonable people and know that we shouldn't overeat. We have done it before, wished we hadn't, and vowed to never do it again. Nonetheless, time after time, we repeat the same mistakes. Are we weak-willed, morally corrupt, and self-destructive? Do we need years of therapy? The answer is "none of the above."

The answer is in our genetic programming. This stuff is just too deep inside us to get away from. We are built to put on weight, and our bodies don't like it very much when we don't give them the calories they need.

To make matters worse, when you lose weight, only about half of what is lost is fat; the rest is valuable, metabolically active muscle! Yet when weight is regained, it is nearly all fat. Remember, muscle cells burn seventy times more calories than fat cells. Therefore yo-yo dieting makes you lose a big part of your metabolic engine. If you follow all the steps of the UltraMetabolism Prescription, you will minimize the loss of muscle and keep your metabolic fire going even while you lose weight.

We all know overweight people who say, "I don't really eat that much, and I still can't lose weight." They aren't lying.

When most people go on a diet, they are generally actually making themselves fatter. Each time they diet, they lose muscle. The diet usually fails, and when it does, the weight that is regained is fat. If you have been through a number of diets that have failed, your body has been through this process a number of times. In short, dieting makes you fat.

Luckily, UltraMetabolism isn't so much a diet as a way of eating. This means that you aren't in danger of falling into this trap if you follow the program in this book.

Summary

- Losing weight is about more than just eating less and exercising more.
- It is not your fault that you are struggling with your weight,

because your feeding behavior is controlled by your lizard brain, an ancient survival mechanism designed to prevent you from starving.

❖ You have to eat more than your resting metabolic rate, or your body will think you are starving. When you eat less than your RMR or resting metabolic rate, you tend to gain weight rather than lose it.

❖ Don't "diet"—if you eat too few calories, you will only gain weight in the long run.

CHAPTER 2

THE CALORIE MYTH:

All Calories Are Created Equal

The Dangers of Counting Calories

*S*andra was very diligent and fastidious about weight loss. A 46-year-old corporate lawyer, she wanted to look trim and fit, not to mention the hassle and expense of buying new clothes when her weight fluctuated. She was also a little bit obsessive-compulsive. She always carried a pocket calorie counter, and she measured and analyzed everything she ate.

Fad diets, she felt, were for those who were not serious about weight loss, and she was quite intent on following the government eating guidelines contained in the USDA food pyramid, which encouraged her to eat six to eleven servings (1 serving = ½ cup) of bread, cereal, rice, and pasta a day. Reassured by scientific evidence and the endless advice of diet experts that if she ate fewer calories, regardless of their origin, she would lose weight, she counted and measured everything.

She didn't overeat, nor did she eat sweets or junk food. But she did use convenience foods because of her busy legal career. Even though they may have been high in ingredients such as trans fats or high-fructose corn syrup, she felt she was fine as long as she stayed within her calorie limit.

But this strategy didn't work for her. She found it increasingly difficult to lose weight, always felt hungry and tired, and was getting fed up with measuring and counting everything. When we looked at her diet carefully, it became clear that she ate a lot of "empty calories"—refined carbohydrates and toxic fats of little nutritional value—and was sending all the wrong messages to her genes, turning on the messages that promoted weight gain, changes in hormones, increases in molecules that trigger more hunger, inflammation, and more—all the things that gave the wrong information to the body as it tried to regulate its weight.

I got her to throw away her calorie counter and guided her in choosing whole foods that contained no toxic or hydrogenated fats or high-fructose corn syrup, and weaned her from convenience foods by helping her with a little planning and shopping so she always had healthy choices. I even had her pack a small cooler in the car so she didn't have to rely on quick fixes of poor-quality calories. Very quickly her hunger cravings stopped and felt liberated from the counting and measuring. And

weight steadily came off and stayed off without any effort other than a little preplan-
ning and shopping.

> The concept that "a calorie is a calorie" underlies most conventional
> weight loss strategies. According to this principle, obesity results from an
> imbalance between energy intake and expenditure. The proposed cure is
> to eat less and exercise more. However, calorie-restricted, low-fat diets
> have poor long-term effectiveness in the outpatient setting. In a sense,
> these diets may constitute symptomatic treatment that does not address
> the physiological drives to overeat. From a hormonal standpoint, all calo-
> ries are not alike.

—DAVID LUDWIG, M.D.[1]

Why All Calories Are *Not* Created Equal

In the last chapter you learned that calorie restriction doesn't work. That's
the reason most diets fail. Dropping extra pounds is not a simple matter of
eating less. In fact, if you eat too little you set off a chain of molecular events
inside your body that will actually cause you to gain weight.

Nonetheless, our culture tries to convince us that if we consume fewer
calories we will lose weight. But this doesn't tell the full story. Calories are
important. But it isn't the *amount* of calories as much as the *type* of calories
you consume that makes a difference in terms of how much you weigh and
how healthy you are. But before we explore that, let's have a closer look at
what calories are and what they do.

What Are Calories, and Where Do They Come From?

What is a calorie, anyway? A calorie is simply a unit of energy. It is defined
as the quantity of energy required to raise the temperature of 1 gram of
water by 1 degree Centigrade at sea-level atmospheric pressure.

We get calories from the food we eat. We consume food, and the chemical processes that make up our metabolism break down this food and turn
it into energy. Burning this energy created by our metabolic machine allows us to do everything from breathing to running marathons.

It's sort of like putting fuel into a car: you have to put fuel in to make the
car run. This is exactly the case with people and food. Food is our fuel. We
consume calories so that we will have something to burn. It is these fuel
calories that make us run.

We need a certain amount of caloric intake just to keep the basic functions of our body operating. And then we need some additional calories to do things such as get out bed in the morning or go for a run. You learned about all this in the last chapter.

A few hundred years ago, Isaac Newton proved that all energy in the universe is conserved; this is known as the first law of thermodynamics. Applied to your weight and what you know about caloric intake, this law suggests that if you eat the same number of calories you burn, you will stay the same weight. If you eat more calories than you burn, you will gain weight; if you eat less than you burn, you will lose weight. This seems perfectly logical. The problem? It's not true.

A Lesson from Physics

I certainly wouldn't presume to throw out Isaac Newton's laws. But how do they actually apply to the calories you eat?

Let's examine a similar physics-oriented example that you probably remember from high school. Take one pound of feathers and one pound of lead and drop them in a vacuum. Which drops faster? Those of you who answered "the lead" need a refresher course in physics.

In a vacuum they both drop at the same rate. They are the same weight—one pound—and they have the same mass, so they drop at the same rate. Now take that same pound of feathers and pound of lead and drop them off the George Washington Bridge in New York City. Which drops faster? If you answered "the lead" this time, congratulations.

Why? Air resistance. You can't see it, you can't taste it, and you can't smell it. But air resistance is real, and it affects how lead and feathers move through it. Even though the lead and feathers in this example have the same weight/mass, they have different properties that cause them to move through air differently.

Calories behave in the same way. When calories are burned in a laboratory, they are all created equal and release the same amount of energy. There is no difference between a thousand calories of kidney beans and a thousand calories of a low-fat muffin or cola—*until* they are metabolized.

Your body's metabolism is like the air resistance in the example above. The calories you eat are absorbed at different rates and have different amounts of fiber, carbohydrates, protein, fat, and nutrients—all of which translate into different complex metabolic signals that control your weight. You may not be able to see, taste, or smell your metabolism any more than you can see air resistance, but it has an impact on how calories are consumed in your body just the same.

For example, the sugar from a soda enters your blood very rapidly, while the same amount of sugar from kidney beans enters your blood slowly. If you drink a soda and all the sugar in it goes into your bloodstream at once, the calories you aren't using at that moment will be stored as fat. On the other hand, if you eat the kidney beans and the sugar in them is absorbed over time, your body has a greater chance to make use of those calories. That means more of them will be burned and less will be stored. Also, because of the high fiber content of the beans, not all the calories will be absorbed.

Food that enters your bloodstream quickly promotes weight gain; food that enters slowly promotes weight loss.

Recent studies have turned the idea that all calories are created equal on its head. Studies show that high-carb diets made up of rapidly absorbed sugars can increase blood sugar and insulin levels, causing weight gain, as well as increasing cholesterol and triglycerides, which lead to a fatty liver, in turn causing even more weight gain.

In a recent study, leading nutrition researchers, including Walter Willett, M.D., and his group from the Harvard School of Public Health[2] designed a study to determine whether low-fat or low-carbohydrate diets were better for losing weight.

The results were startling. For twelve weeks, the researchers fed their group of overweight patients three different diets, all carefully controlled and prepared for them daily in a Boston restaurant. The first group ate a low-fat diet of 1,500 calories (55 percent carbohydrate, 30 percent fat, 15 percent protein) for the women and 1,800 calories for the men. The second group ate the exact same number of calories but from a low-carbohydrate diet (5 percent carbohydrate, 30 percent protein, 65 percent fat). The third group also consumed a low-carbohydrate diet, but they ate 300 *more* calories a day than the other group: 1,800 for women and 2,100 for men.

The researchers discovered that the low-carb group eating the same number of calories as the low-fat group actually lost more weight. The low-carb group lost an average of 23 pounds, compared to 17 pounds for the low-fat group, despite eating exactly the same number of calories. That's 6 pounds more in 12 weeks. While the study was only 12 weeks long, the findings are worth noting.

The real question is what type of carbs and fats were used. The low-carb diet was predominantly a diet of whole, unprocessed foods—lean animal protein, vegetables, whole grains, and beans—in other words, a basic Mediterranean-style diet. The low-fat group ate foods that were higher in refined carbohydrates. But as we will learn here, the low-carb movement will also eventually go by the "weigh-side."

What was more startling was that the group eating 300 more calories a day with the low-carb diet lost more weight than those eating the low-fat diet, even though the 25,000 more calories they ate should have amounted to 7 pounds of increased weight. They actually lost an average of 20 pounds, or 3 pounds per person, more than the low-fat group, who ate 25,000 fewer calories during the 12 weeks.

One final study drives the point home. A Harvard professor, Dr. David Ludwig, studied three groups of overweight children,[3] feeding each group a breakfast containing an identical number of calories. One group ate instant oatmeal; one group ate steel-cut oats (the type that takes about 45 minutes to cook); and the third group had a vegetable omelet and fruit.

Their blood was measured before they ate and every 30 minutes afterward for the next five hours. Then they ate a lunch identical to the meal they had eaten for breakfast. After finishing lunch, they were told to eat whenever they were very hungry for the rest of the afternoon. What happened was startling.

Many of you would think that the healthiest breakfast would have been the oatmeal. But it was actually the omelet. The group that ate the instant oatmeal (the breakfast that entered the bloodstream and turned to sugar the fastest) ate 81 percent more food in the afternoon than the group that had the omelet. Not only were they hungrier, but their blood tests looked entirely different. The instant oatmeal group had higher levels of insulin, blood sugar, blood fats, and adrenaline even though they consumed the same number of calories as the omelet group. Though the steel-cut oats were better than the refined oats, the children who ate the steel-cut oats still ate 51 percent more food in the afternoon than the children who ate the omelet. (Adding nuts, soy milk, and flaxseeds to steel-cut oats allows the meal to be absorbed more slowly.)

The conclusion here is that the kinds of calories you consume have a big impact on how much weight you gain, because different types of food are metabolized in different ways.

What's even more interesting is the fact that the type of calories actually has an effect on how your metabolism functions. The type of food you eat has a big impact on what your genes tell your metabolism to do. This means that the types of calories you consume have a dual impact on the way you metabolize food. They act as both a source of energy *and* a source of information or instructions to your genes that control metabolism. Let's take a closer look at the way food talks to your body.

Food "Talks" to Your Genes, and Your Genes "Talk" to Your Body

We used to think the human genetic code—DNA—was simply a set of data that dictated things such as what color your eyes are, how tall you are, and what you look like. The old assumption was that this code simply sat in storage somewhere in your cells until it was passed on to your children. The genomic revolution has opened a whole new world of understanding about what our genes really do.

Your genes do control your physical characteristics to some degree. But that is only a fraction of their job. They actually control the day-to-day flow of instructions that regulates every aspect of your biochemistry and physiology. They control the production of hormones, brain messenger chemicals, blood pressure, and cholesterol, as well as mood and aging processes, and they even play a role in your risk of acquiring diseases like cancer and heart disease. Essentially, they control every function of your body from moment to moment. Your genes play an especially important role in controlling your metabolism and your weight.

What's more, nutrigenomics has revolutionized our understanding of food and calories. We have recently discovered that food is more than just energy or calories.

Food contains hidden information. This information is communicated to your genes, giving your metabolism specific instructions on what it should be doing. Some of the instructions food gives are: lose weight or gain weight; speed up or slow down the aging process; increase or decrease your cholesterol level; produce molecules that increase or decrease your appetite. The kind of food you eat gives your genes different information, helping them make decisions as to what it will tell your body to do in these and various other areas. *Food talks to your genes.*[4]

The new science of nutrigenomics teaches us what specific foods tell your genes. What you eat directly determines the genetic messages your body receives. These messages, in turn, control all the molecules that constitute your metabolism: the molecules that tell your body to burn calories or store them.

If you can learn the language of your genes and control the messages and instructions they give your body and your metabolism, you can radically alter how food interacts with your body, lose weight, and optimize your health. You can either learn to speak this language or suffer the consequences of serious miscommunication: weight gain, fatigue, and disease.

Teaching you to speak the language of your genes is what the UltraMe-tabolism approach is all about.

Living in Harmony with Our Genes:
Eating a Whole-Food Diet

We need to eat in harmony with our genes. Because each of us starts with a different set of DNA, living harmoniously with our genes will mean something different to everyone.

Some of us need more fat, protein, or carbohydrates than others. There is no one perfect diet for everyone. You need to find out what works for you. But your metabolism has some basic operating principles that we all share, and there are specific tests and clues to discover what affects them.

Looking at each of these principles is what part II of this book will help you do. However, there is one principle I would like to introduce here, because it is such a fundamental part of the UltraMetabolism program. That is the importance of eating whole, unprocessed foods.

A whole, unprocessed, real food is one that is as close to its natural state as possible when you buy it at the grocery store—a whole avocado, a whole apple, a whole grain, a whole almond, or a whole tomato. Almost anything made or packaged in a factory (i.e., anything with a label) is *not* a whole food.

Whole foods evolved with humankind over thousands of generations. Our bodies adapted to them, and they adapted to our bodies. The calories you consume that come from whole foods speak to your genes in their native tongue. Your DNA knows exactly what to tell your metabolism to do to use these foods in the most efficient and healthy manner possible.

Whole foods are not tainted with unhealthy fats and refined carbs (you will learn more about this in chapters 3 and 4) or manmade elements that your body has no idea how to process properly. Whole foods were designed by nature to keep you at a healthy weight and give you UltraMetabolism.

This doesn't mean that you need to go out today and start stocking your shelves exclusively with organic items, although there is a value to buying organic foods that we will discuss later. Nor does it mean that you need to immediately eliminate all the packaged goods from your kitchen cupboards, although you will see in part III of this book that a little bit of pantry preparation will help your diet. What it does mean is that you need to focus on making whole foods a substantial part of your diet. If there is one thing I would recommend you start doing right now, it is this.

We will be discussing whole foods throughout the rest of the book, and

WHAT IS A WHOLE FOOD?

Whole foods are foods that are in the form found in nature—fresh, unprocessed, and simple. These include:

High-fiber foods

- Beans
- Whole grains
- Vegetables
- Fruits
- Nuts
- Seeds

Quality Proteins

- Beans
- Nuts and seeds
- Eggs
- Fish
- Lean poultry, lamb, pork, or beef (preferably organic, grass or range-fed)

Healthy Fats

- Fish oil
- Extra virgin olive oil
- Cold-expeller-pressed plant oils, such as grapeseed, walnut, and sesame
- Avocado
- Olives
- Coconut
- Nuts
- Seeds

Healthy Carbohydrates

- Vegetables
- Whole grains
- Beans
- Fruit
- Nuts
- Seeds

by the end you should have a very good idea of what they are. For now, see the sidebar for additional information that you can start using immediately.

Note that whole foods often contain a mixture of protein, fat, carbohydrates, and fiber, so you will find many whole foods in multiple categories.

Summary

- ✢ All calories are not created equal.
- ✢ Food contains information for your genes that controls your metabolism, not just energy in the form of calories.
- ✢ The information changes your metabolism and can increase or decrease your weight.
- ✢ A whole-food diet is the best way to help food communicate to your genes in a language it understands.

CHAPTER 3

THE FAT MYTH:

Eating Fat Makes You Fat

When Fat Makes You Thin

*W*hen Paul first saw me, he was desperate. At 42 years of age, he had had his first heart bypass. Terrified he would die before he was 50 years old, he followed his doctor's advice perfectly. He ate a perfect low-fat diet, exercised daily, took an aspirin, a cholesterol-lowering medication, and a beta-blocker to protect his heart.

After eight years he once again developed chest pain and went to see his doctor. Another angiogram showed that his new "bypass" arteries had clogged. The doctor performed a procedure called an angioplasty, in which a balloon is inflated inside the artery to correct the blockage. Again his doctor advised him to eat even less fat. Worried for his life, he reduced his fat intake further.

Over the next year he had six more angioplasties, and after each one the doctor advised further reduction in his fat intake. He gained weight after each angioplasty. Don't worry, his doctor told him; as long as his fat intake was low he would be fine. He complained about the weight gain each time, and each time he was reassured. Finally he needed another bypass.

By the time he saw me he was tired, 30 pounds overweight, having night sweats, and feeling terrible. I took one look at him and told him that his low-fat diet of pasta, rice, bagels, and potatoes was killing him. Without ever seeing his blood tests but simply looking at his round belly and hearing his story, I told him he had a low cholesterol level but a high triglyceride level and a low HDL, or "good," cholesterol level as a result of prediabetes or insulin resistance. I had him cut out the high-glycemic-load carbs (quickly absorbed refined carbs), eat more nutrient-rich whole foods, including vegetables, fruits, and beans, and add foods containing healthy fats, including olive oil, nuts, seeds, avocados, coconut oil, and fish oil to his diet.

Within a few months he had lost 30 pounds, his night sweats were gone, his energy was better than ever, and all his blood work was normal. Despite his growing belly, both he and his doctor believed so strongly that fat makes you fat that they missed the obvious.

Mainstream nutritional science has demonized dietary fat, yet 50 years and hundreds of millions of dollars of research have failed to prove that eating a low-fat diet will help you live longer.

—GARY TAUBES[1]

Diets high in fat do not appear to be the primary cause of the high prevalence of excess body fat in our society, and reductions in fat will not be a solution.

—WALTER WILLETT, M.D., PH.D.[2]
Professor, Harvard School of Public Health

We have been brainwashed to believe that if we eat fat, we will get fat. We hear these messages everywhere we go: "Eat fat, and you will gain weight; avoid fat, and you will lose weight."

The U.S. government (Department of Health and Human Services, 1988), the American Heart Association (1996), and the American Diabetes Association (1997) have all recommended a low-fat diet to prevent and treat obesity. It seems perfectly logical: if you don't eat fat, you won't gain fat. There's one problem—science does not support this recommendation.

We have been hoodwinked by shaky science and political lobbying into believing that if we eat fat we will get fat because fat has more calories per gram than carbs (9 versus 4). But there is more to the story than calories or fat grams.

The American Paradox: We Eat Less Fat, We Gain More Weight

Recently Dr. Walter Willett from Harvard and others have shown that high levels of dietary fat do *not* promote weight gain, and any weight loss from a low-fat diet is usually modest and temporary. Two studies published in *The New England Journal of Medicine*[3] found that a low-carbohydrate diet leads to more weight loss than a low-fat diet.

Over the last thirty years, Americans have tried to get healthy by cutting fat out of their diet. But our obesity rate has tripled since 1960, and a full two thirds of our population is presently overweight. We now have an epidemic of obesity and diabetes in children. **Because of the obesity epidemic, for the first time in the history of the human species life expectancy is declining, not increasing, so that children of this generation will live sicker and die younger than their parents.**[4]

In the last forty years, our national fat consumption has decreased from

42 percent to 34 percent of our total calories. We're eating less fat than we ever have (as a percent of total calories), but we're growing fatter. This is the American Paradox.

One of the major reasons this is the case is that low-fat diets are often rich in starchy or sugary carbohydrates. In our attempts to avoid fat, we began to replace "fatty foods" with easy-to-access carbohydrates (white flour, white rice, pasta, potatoes, and sugars). These carbohydrates help temporarily replace the feelings of hunger that are left by a lack of fat in the diet. In addition, they are incredibly easy to produce and distribute, so the food industry has invested a great deal of money and energy in doing so. Now they are nearly ubiquitous in our culture. These types of carbohydrates raise insulin levels, which, in turn, promote weight gain. It is also easier to eat a lot more of them because while fat makes you full, sugar just makes you hungrier. You will learn more about this and other carbohydrate myths in chapter 4.

There Is No Evidence That Fat Is Bad for You

The great irony in all this is the fact that there is absolutely no scientific evidence to support the idea that a low-fat diet contributes to either weight loss or good health. You may find that a bit hard to swallow, because the idea is so prominent in our culture. Nonetheless, it's true. Have a look at the following studies.

Consider the data on the relationship between high-fat diets and heart disease. We all "know" that a high-fat diet causes heart disease. We have been taught that eating fat increases our cholesterol and high cholesterol causes heart attacks. From this information we conclude that decreasing the fat in our diet would lead to fewer heart attacks. But while death from heart disease seems to be dropping, the number of people acquiring heart disease is not. The American Heart Association's own statistics show that between 1979 and 1996, medical procedures for heart disease increased from 1.2 million to 5.4 million a year. Heart disease isn't decreasing with the low-fat diet America has adopted; we are just better at treating heart disease once it exists.

Another famous project, the Lyon Heart Study,[5] had to be stopped prematurely because people eating the low-fat American Heart Association diet were dying, while those eating the healthy higher-fat Mediterranean diet, including olive oil, olives, nuts, avocadoes, and fish, were doing fine.

A more recent study[6] found that over ten years healthful lifestyle practices in an older population (70- to 90-year-olds), including a higher-fat, Mediterranean diet, moderate physical activity, nonsmoking status, and moderate alcohol consumption, were associated with nearly a 70% reduction in death from *any* cause.

An example that is almost too absurd to believe is the Harvard Nurses' Health Study, which involved over 300,000 women studied over a ten-year period, to find out if there were any correlation between dietary fat and heart disease. The U.S. government spent nearly $100 million on the study, hoping to prove that dietary fat was indeed a killer.

The study ultimately found that there was no connection between the two, but the government refused to change public policy, which had been built on the idea that a low-fat diet was a healthier way to eat years before the study was completed. Dr. Willett, the lead researcher and spokesperson for the project, even publicly decried the government's reaction as "scandalous," but there was no effect. Public policy for a low-fat diet is still on the books to this day, even though it was adopted without a shred of scientific evidence.

The truly unfortunate part about this scandalous decision is the fact that U.S. policy on low-fat diets has contributed to an epidemic of obesity, diabetes, heart disease, and even cancer-related disorders. How? By encouraging Americans to adopt a low-fat diet and recommending in the 1992 USDA Food Pyramid that it's healthy to eat six to eleven servings of cereal, rice, bread, and pasta in place of fat. This pattern of eating has been proven to contribute to every one of these fatal health conditions.

How Pharmaceutical Companies and the Food Industry Are Conspiring to Keep You Fat

To make matters worse, the public's belief in the mythology that surrounds dietary fat has been supported for years by some of the most powerful corporate entities in this country: the pharmaceutical and food industries. The support they give to a low-fat diet isn't founded on scientific data any more than the medical industry's is. And the unfortunate truth is that there is a great deal of evidence to suggest that they have promoted these unfounded, unhealthy beliefs simply for financial gain.

Drug companies would have you believe that "bad," or LDL (low-density-lipoprotein) cholesterol, is the single biggest factor in the development of heart disease. But the truth is that the determining factor is the ratio of your *total* cholesterol to your "good," or HDL (high-density-lipoprotein), cholesterol. (There is an easy way to remember which cholesterol is good and which is bad: the "L" in LDL is for "lousy," and the "H" in HDL is for "happy.")

The pharmaceutical industry promotes this unfounded belief not because there is scientific evidence that supports it but because the main class of drugs available for treating high cholesterol are statins, which mostly lower LDL, and these drugs are among the biggest-selling drugs in history.

The truth is that the ratio between your total cholesterol and your HDL is almost entirely determined by the type and amount of carbohydrates you eat, *not* by the amount of fat you consume.[7]

The chart below illustrates this point nicely. All the items below the mean value line improve the ratio between your total cholesterol and your HDL. Everything above the line puts it out of balance. If you look at the chart, you will see that carbohydrates are by far the greatest culprit in worsening the ratio between your total cholesterol level and your HDL level (see Figure 1).

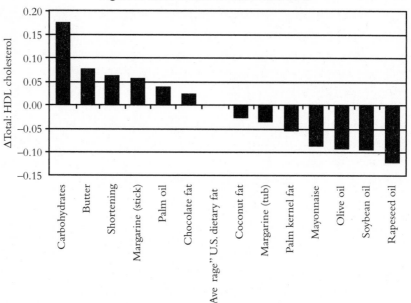

Figure 1: Ratio of Total to HDL Cholesterol

"Bad" carbohydrates are a staple of the food industry and the low-fat diet. Shifting from refined carbohydrates such as bread, pasta, rice, and sugar in all its forms to "good" (healthy) carbohydrates such as vegetables, beans, whole grains, and fruit would lead to a dramatic reduction in all the diseases of aging and obesity. But the food industry spends $30 billion a year on advertising to convince us to eat foods that include bad carbohydrates, such as fast food, sodas, snacks, and candy bars. Indeed, can you imagine the loss of profit such companies would experience if they changed their menus from supersized French fries to fruit? The food industry makes up 12 percent of our gross national product (GNP), so it is both a health issue and an economic one.

There is only one way to overcome the scandal that surrounds fat in this country: you need to develop a better understanding of what fats do and what they tell your metabolism, and keep in mind that all fats are not created equal.

All Fats Are Not Created Equal

So are all fats created equal? Fat chance! Some fats are good for you and actually help you lose weight; others make you put it on. As you know from chapter 2, different kinds of food interact with your body in different ways. This holds true for fat. There are certain types of fat that are healthy and certain types that are lethal. The problem is that most of the healthy kinds of fat have been eliminated from our diet and most of the fats that are lethal are so predominant in the modern-day food supply that they are hard to avoid.

Whether or not a particular fat is healthy or unhealthy depends largely on the kind of information the fat communicates to your genes. "Good" fats communicate messages of health and weight loss; "bad" fats communicate unhealthy messages that contribute to weight gain.

To communicate these various messages, molecules from fat cells bind to special receptors on the nucleus of your cells called *PPARs* (peroxisome proliferator activated receptor).★[8]

Different types of fat interact with your PPAR receptors in different ways. Bad fats (see the next page for more information on different fats) turn off your fat-burning genes, making it much harder for you to lose weight. On the other hand, when you consume good fats, they bind to the same PPAR receptor but turn on genes that increase your metabolism, help you burn fat, and make you more insulin-sensitive.[9] (Insulin resistance is a condition where you develop a tolerance to insulin and produce excess amounts of insulin. Generally this is because you eat too many sugars or bad carbs. Doing so is a major contributor to every known degenerative health condition. You want to be insulin-sensitive, not insulin-resistant. There will be more information on insulin resistance and sensitivity in the next chapter.)

For example, studies have shown that the fat in fish oil, EPA (a good fat), binds to the PPAR receptor, turning on fat burning and improving insulin sensitivity,[10] while trans fats (really bad fats) actually have the exact opposite effect, blocking your metabolism and slowing down fat burning.[11]

So here's the take-home message in all this: eat the right fats, and you turn up your fat-burning capacity; eat the wrong fats, and you switch on

---

★ PPARs are a new class of nuclear receptor that is critical in controlling insulin sensitivity, fatty acid oxidation or fat burning, and inflammation.

genes that make you gain weight and slow down your metabolism. The *type* of fat you eat is more important than the *amount* of fat you eat.

The trick is knowing what kinds of fats are "good" and which ones are "bad." In the section below I will give you a brief outline of different kinds of fats and whether or not they are healthy or unhealthy.

Understanding Fats: The Good, The Bad, and the Ugly

There are essentially three different kinds of fat: the good, the bad, and the ugly. "Good" fats are fats that turn on the genes in your DNA that increase your metabolism, help you burn fat more quickly, and become healthier. "Bad" fats are the ones that affect your metabolism adversely, making it difficult to burn the weight you would like to.

"Ugly" fats are a different animal entirely. These are man-made fats that simply cannot be properly digested by your body at all. These fats interrupt the natural operation of your cells and have the capacity to affect your health in radically negative ways.

In each of these categories I will give you the name of the fat, a brief description of how it interacts with the body, and some examples of places where you might find this fat.

Good Fats

Omega-3 Fats

The king of the good fats is the omega-3 fats. These are fats that come from wild foods. We evolved eating these "essential" fats because as recently as 10,000 years ago, before the agricultural revolution (a split second in evolutionary time), we were *all* eating wild foods.

The problem is that 99 percent of us are currently deficient in these healthy, essential fats. Despite their critical importance in our diet, most of us don't eat enough unprocessed, wild, whole foods anymore, with the potential exception of fish. Unfortunately, most of the fish we consume is farmed or from oceans polluted with mercury and hence full of PCBs or heavy metals—both of which are highly toxic to your body.

Today you can find these fats in a few whole, unprocessed foods. The following list shows examples:

- Wild fish, including wild salmon, herring, sardines, and fresh anchovies (avoid nonorganically farmed fish and large predatory fish such as tuna and swordfish, which accumulate more mercury)
- Flaxseeds and flax oil

❖ Some types of nuts and seeds, including walnuts, pumpkin seeds, and hemp seeds

Monounsaturated Fats

Mediterranean diets provide up to 40 percent of their calories from fat, mostly monounsaturated fats derived predominately from olive oil. This type of diet reduces the risk of chronic diseases such as heart disease, diabetes, and cancer. Olive oil intake has been shown to reduce inflammation, boost immunity, and contains powerful plant antioxidants called phenols. It reduces blood pressure and blood sugar, lowers cholesterol, and thins the blood. Monounsaturated fat is considered to be among the healthiest types of fat. It has none of the adverse effects associated with saturated fats, trans fats, or omega-6 polyunsaturated vegetable oils. The best sources of monounsaturated fat are:

❖ Extra virgin olive oil (73 percent)
❖ Hazelnuts (50 percent)
❖ Almonds (35 percent)
❖ Brazil nuts (26 percent)
❖ Cashews (28 percent)
❖ Avocado (12 percent)
❖ Sesame seeds (20 percent)
❖ Pumpkin seeds (16 percent)

Some Saturated Fats

Although many saturated fats are bad fats (such as dairy fat or myristic acid), there are some that are actually good for you. Most of the negative focus on saturated fats comes from research about their effects on LDL cholesterol, and dairy fat in particular has the worst effect.

However, human mammary glands produce many saturated fats that must be necessary for the growth and development of a young child. The human brain is actually made up of 60 percent fat, including a special saturated fat called lauric acid and omega-3 fats. These fats are also found in our cell membranes and are used as preferred sources of energy by heart cells.

Unfortunately, little research has been done on the various types of saturated fats. It is best to include some saturated fat in your diet, ideally from sources high in lauric acid such as coconut products, including coconut, coconut milk, and coconut oil. But ultimately you should keep it to a minimum in your diet (less than 5 percent of total calories) by reducing common sources, such as feedlot or commercially raised beef, pork, lamb, and poultry. Range, grass-fed, grass-finished, and organically fed animals typically have much less saturated fat in their tissues. A feedlot-raised steer, for example, has 500 percent more saturated fat in its tissues than a grass-fed steer.

Some of the "better" saturated fats include those in coconut oil, which contains lauric acid, a saturated fat found in mother's milk that boosts immune function and helps the body kill viruses and yeasts. Recommended forms of coconut and better forms of saturated fat include:

- Raw coconut
- Coconut milk
- Coconut oil
- Palm fruit oil
- Macadamia nut oil

Unrefined Omega-6 Polyunsaturated Fats (in small doses)

In small doses, certain types of unrefined omega-6 polyunsaturated fats are necessary. These are natural vegetable oils that have not been chemically processed. The problem in today's world is that the ratio of omega-3 to omega-6 fats in our diet has changed dramatically. We have been flooded with polyunsaturated vegetable oils of poor quality. These "refined" oils include most commercially available cooking oils: corn oil, "vegetable oil," safflower oil, and others. Check your own kitchen cabinet and see if you have a bottle of this stuff in there. It is almost everywhere these days, and the amount in which it is used is very unhealthy.

Nonetheless, use of expeller or cold-pressed versions of these oils in small amounts is necessary, because our bodies evolved with a balance of beta omega-6 and omega-3 oils. A few examples of these fats that you can use in small doses and still maintain optimal weight loss are:

- Grape seed oil
- Sunflower oil
- Safflower oil
- Walnut oil
- Sesame oil

Bad Fats

Refined Polyunsaturated Vegetable Oil

These include most commercially available vegetable oils—corn, soy, safflower, and "vegetable oil." See the description of unrefined omega-6 polyunsaturated fats above for a further discussion.

Most Saturated Fats

Beef, pork, lamb, chicken, and dairy foods are the main sources of saturated fats in our diet. It is a common misperception that eggs contain a lot of saturated fat. The average egg has only 2 grams and the newer omega-3 eggs have even less. Eggs contain cholesterol, however, the amount of cholesterol in food is so minimal that it has no significant effect on your blood cholesterol. Eggs are a good source of protein if you are not allergic to them. It is even better if you can buy omega-3 eggs, which are a safe source of the important essential fats. Shellfish also has cholesterol but minimal saturated fat.

Remember, eating cholesterol does not raise your cholesterol. Nearly all the cholesterol in your blood is manufactured from saturated fats and sugars or refined carbohydrates. It is important to note that you are what you eat, even if you are a cow or a pig. As I mentioned on page 36, feedlot-raised beef contains 500 percent more saturated fat than grass-fed beef. There are many health and weight reasons to choose "cleaner" forms of animal protein, such as a reduced intake of pesticides, hormones, and antibiotics, as well as a reduced risk of prion diseases such as mad cow. I recommend doing your best to find better-quality food whenever possible, though these foods are sometimes more expensive and difficult to come by (in the resources section I have listed some good sources of "clean" animal products).

Ugly Fats

Hydrogenated Oils or Trans Fats

Most dangerous of all are the fake "trans fats," or partially hydrogenated fats.* These relatively new man-made fats (margarine or shortening are the primary examples) were originally developed to manufacture fake butter from vegetable oil during a butter shortage. This seems like a nice idea, but there is one problem: they're toxic. They block your metabolism, create weight gain, and increase your risk of diabetes, heart disease, and cancer.

These fake fats have no place in the diet of any known species on the

---

* Trans fat occurs in foods when manufacturers use hydrogenation, a process in which hydrogen is added to vegetable oil to turn the oil into a more solid fat. Trans fat is often found in vegetable shortening, some margarines, crackers, candies, cookies, snack foods, fried foods, baked goods, salad dressings, and other processed foods.

planet. Yet we all consume them in large quantities without the slightest awareness of what we are doing. They are found in nearly every processed or commercially baked or packaged food because they never spoil. They are like plastic: they stay stable in landfills for generation after generation. Ever wonder how a package of Ritz crackers can have an expiration date of a few years from now? Keep a tub of margarine open on the counter, and you'll find that no bug will go near it!

Despite the opposition of powerful food lobbies, the FDA finally mandated the labeling of foods containing trans fats, beginning on January 1, 2006. This was almost three years after it created the regulation in July 2003 and more than a decade after the dangers of these fats were clear. The FDA estimates that just making people aware of the dangers of trans fats with the new labeling regulations will save between $900 million and $1.8 billion each year in medical costs, lost productivity, pain, and suffering.

Why are trans fats so bad? There are many reasons, only some of which we understand completely. Clearly, they seem to disrupt our metabolism. If you eat Twinkies or Hostess Cupcakes (my childhood favorite), you are consuming large amounts of trans fats. Those fats speak directly to your DNA, turning on a gene that slows down your metabolism, causing you to gain weight.

These days you find these terrible, ugly fats in so many places it would be hard to make a comprehensive list of them. They are in virtually *all* processed foods. But they can sometimes come with slightly sneaky names attached: "hydrogenated" and "partially hydrogenated" oils are the two big culprits. Read food labels, and look out for those terms, which mean there are trans fats buried within.

And If You Still Don't Buy What I'm Telling You . . .

If you still underestimate the importance of fat in your diet, consider the following example.

The Greenland Inuit population, which has been studied extensively, used to consume a diet of 70 percent fat, an extraordinary amount of fat by modern standards, yet were thin and free of heart disease, obesity, and diabetes. Why? They ate polar bear, seal, artic char, whale, walrus, and other easy-to-find foods (if you live at the North Pole!). The fat in these foods was predominantly omega-3 fats and monounsaturated fats.

The Inuit were a fit race, well adapted to its environment—until recently. In modern times the Inuit shifted from a diet high in good fat to a diet low in fat and high in processed carbohydrates. For reasons you will learn more about in the next chapter, this made the population incredibly obese in very little time.

The take-home message: a diet high in fat is not a bad thing. It is the *type of fat* that is important.

Fat and the UltraMetabolism Prescription

Learning that fat is not a bad thing is an important step in healing your metabolism. It is important to incorporate good fats (especially omega-3 fatty acids from wild fish and nuts) into your diet. A low-fat diet will not make you lose weight. A diet high in "bad" fats is even worse.

The UltraMetabolism Prescription will show you a way to keep good fat in your diet so you can turn on the genes that will make you lose weight and watch the pounds fall away.

Summary

- Low-fat diets are a myth this culture has promoted for too long. They don't work and have no scientific evidence to support them.
- It is not a question of eating a low-fat versus a high-fat diet; it's a question of the type of fat you eat. As you can see, your diet is a powerful messenger that communicates to your metabolism through the messages the fats send via your genes.
- Essential omega-3 fats are good fats. They come from fish, flaxseeds, and nuts, enhance and improve your metabolism, and promote weight loss. Eat more of them.
- Trans fats are ugly fats that cause weight gain, impaired metabolism, inflammation, and diabetes. *Never* eat them.

CHAPTER 4

THE CARB MYTH:

Eating Low Carb or No Carb Will Make You Thin

Clearing Up the Carb Confusion

*J*onathan, a 36-year-old insurance salesman, thought he had reached Shangri-La when he heard about a dream diet. Although he did have to give up carbs and bread, he could feast on eggs and sausage, ribs, and a big T-bone steak, put heavy cream in his coffee, and not feel guilty. He could eat as much as he wanted and lose weight—and he did lose weight, about 20 pounds initially.

Though suffering from severe constipation, hemorrhoids, and bad breath, he continued on. But after a while he felt toxic, tired, and actually bored with his bacon-and-cream diet, and his weight stopped coming off. Vegetables and fruit actually started sounding interesting to him.

When I saw him, he was looking for a long-term solution to being overweight, and after his initial infatuation with the all-you-can-eat BBQ buffet, he had realized that eating bacon, cream, and steak every day couldn't be healthy. He was right. Besides increasing his risk of developing cancer, heart disease, kidney failure, and osteoporosis, he just wasn't feeling good anymore. We got him to eat a whole-food diet, including some of his favorite meats (though I recommended he change to range, grass-fed, or organic varieties of these meats), plenty of fiber, vitamins, and minerals (which can quickly become deficient in a person who is on an all-animal-protein-and-fat diet), and slowly absorbed or low-glycemic-load carbohydrates.

His energy perked up, and his hemorrhoids and constipation disappeared along with his bad breath and extra pounds. He had a newfound lifelong strategy for eating that didn't exclude fat or carbs but got him eating the right ones.

Carbohydrates Are the Most Important Foods in Your Diet

If all calories aren't created equal and all fats aren't created equal, there is one question that begs to be asked: Are all carbohydrates created equal? You may be able to guess by now that the answer is no. As with fats, there are *different types* of carbs, and the different types interact differently with your genes, leading to remarkably different effects on your metabolism.

Despite what the American culture at large tells you, *carbohydrates are actually the single most important food in your diet for long-term good health.* Yes, you read that correctly. Without carbohydrates you won't last long. Carbohydrates found in their natural form contain most of the essential nutrients and specialized chemicals that keep you healthy and turn up your metabolism. Unfortunately, human beings have not evolved to metabolize the highly processed carbohydrates so predominant in our current diet. These processed and refined carbs slow down our metabolism and contribute to every one of the major diseases associated with aging, including diabetes, heart disease, dementia, and cancer.

When most Americans think of carbs, they probably think of bread, pasta, and sugar. These things are certainly carbs. But the world of carbs is actually much broader than this.

What Is a Carbohydrate?

Carbohydrates are one of three major energy-producing substances that we consume. The other two are fat and protein. You probably have some idea of what fat and protein are. Fat is . . . fat. Protein is found in both animals and plants. The basic building blocks of life, amino acids, come from protein. The major sources in our current diets are meat, poultry, eggs, dairy, fish, beans, seeds, and nuts.

Carbohydrates are essentially everything else, and they constitute by far the biggest group of foods we eat. Without them, we would starve and die. They make up approximately 90 percent of the living world, and estimates are that 70 to 80 percent of all the calories consumed by human beings the world over are carbs. (It is interesting to note that Americans currently get only about 50 percent of their caloric intake in the form of carbohydrates.)[1]

These days when you say the word "carbohydrate," most people think of what I call "the white menace," namely, white flour, white sugar, or variants on these. Most kinds of bread, pasta, and cereal you find in your local grocery store contain the white menace and are definitely bad carbohydrates for reasons that will be described later in this chapter.

But did you know that vegetables are carbohydrates? So are fruits. Whole grains, beans, nuts, and seeds are also carbs. Each of these is a critical element in the human diet and has been for millennia. If you want to be healthy and thin for the rest of your life, you absolutely *must* eat these good carbs for one major reason: they contain phytonutrients.

Carbohydrates = Phytonutrients

Phytonutrients are healing plant chemicals, and they can be found only in certain types of carbohydrates. The prefix *phyto* simply means "plant," so phytonutrients are nutrients that are found specifically in plant food. The only way you can acquire these important substances is to eat real, whole, unprocessed plant food—all plant foods contain carbohydrates!

Phytonutrients are essential to optimal health. They help turn on genes that make you burn fat and age less quickly. They are the source of nature's most powerful antioxidants and thus they reduce oxidative stress (a concept we will discuss in chapter 12); this in turn reduces inflammation and mitochondrial damage (concepts we will discuss in chapters 11 and 13, respectively), and each of these factors affects your metabolism. For now, suffice it to say that phytonutrients are critical to your health and to a healthy metabolism.

If you have spent any time at all in the world of low-carb dieting, you will have heard of the *glycemic index (GI)*. This index is a concept that is outdated for reasons I will describe more fully below.

The glycemic index should be replaced by an index that judges how rich the carbohydrates you eat are in phytonutrients. This new index is called the *phytonutrient index* (PI), and it gives us much more valuable information than the glycemic index ever did, because it gives us a way to judge how rich our diets are in these healing plant foods. This way of thinking about food has not yet been widely adopted, but it is the simplest way of choosing high-quality food. When you want to eat a certain food, ask yourself if it is a food your ancestors might have eaten. If so, take a bite; if not, put it back!

While there are as yet no good tables or charts to document a food's phytonutrient index, the general concept is extremely important. Another way to think about this concept is to consider the total amount, or *phytonutrient load,* of your diet. Just think of plants in their unadulterated state—fresh, whole, and unprocessed—vegetables, fruits, nuts, beans, seeds, whole grains, and think color and variety. Almost all refined oils, refined sugars, refined grains, potato products, hard liquors, and animal products—regrettably, the chief sources of calories in typical Western diets—have *no* phytonutrients.

From an evolutionary perspective, these phytonutrients are critical parts of our diet. Actually, our bodies are very lazy; if we can give up making something ourselves (we are one of the few mammals that has lost the ability to synthesize vitamin C by ourselves), we do. The phytonutrients in our diet are critical in controlling gene messages that affect both our health and our weight. This is one of the major reasons to eat a predominantly plant-based diet of whole foods. These phytonutrients turn the genes that control weight and metabolism on and off and help prevent every known chronic disease of modern civilization.

More disease-fighting phytonutrients are being discovered in foods every day. Here are some examples: isoflavones in soy foods, lignans in flaxseeds, catechins in green tea, polyphenols in cocoa, glucosinolates in broccoli, carnosol in rosemary, resveratrol in red wine. All of these compounds fight disease and obesity through a variety of mechanisms. They are part of the secret of UltraMetabolism and the key way in which we talk to our genes.

Our ancestors foraged for wild food—wild berries, grasses, roots, and mushrooms. Recently I found myself in a sea of phytonutrients in the wild islands of southeast Alaska, foraging along with the grizzly bears for bog cranberries, blueberries, nagoonberries, raspberries, and strawberries, all smaller, but much richer in color and taste and lower in sugar, than their domestic cousins. Those berries were bursting with phytonutrients. The greater variety and the deeper the color of the plant foods you eat, the higher the concentration of phytonutrients in your diet and the greater their power to prevent disease and promote weight loss.

For example, fresh vegetables score fairly high on the phytonutrient index, while typical pastas and breads don't even make the list. This tells us that fresh vegetables are higher in healing phytonutrients than refined or processed carbs.

When carbs are processed, many of their important phytonutrient properties are stripped from them. This is one of the reasons processed carbs are so bad for you: they are basically empty calories. Ever wonder what they are empty of? They are empty of vitamins, minerals, and phytonutrients. While they increase your sugar and energy intake, they don't offer any of the health benefits provided by whole plant foods rich in phytonutrients.

But processed carbs are bad for other reasons, too. Indeed, good carbs aren't good only because they contain high levels of phytonutrients. As important as phytonutrients are, they tell only part of the carbohydrate story. The types of carbs you eat have an enormous impact on how quickly you metabolize food and how healthy you are, and only part of this is revealed by looking at the phytonutrient index. To complete the picture, we will have a brief discussion about how you metabolize sugar.

Harmonious Sugar Metabolism and How It Turns into Chaos: Insulin and Insulin Resistance (Also Known as Metabolic Syndrome)

Carbohydrates are the source of sugars in our diet—our major source of energy. The company they keep determines the effect of those sugars on our metabolism. If they are all alone, such as in a soda, they are very harmful to our metabolism. If they are found in good company, such as in beans, along with the fiber, vitamins, minerals, and phytonutrients that are always found in whole, unprocessed foods, they have beneficial effects and help us lose weight and keep it off. Sugars found alone in the diet are harmful because they enter your bloodstream quickly, starting a dangerous cascade of molecules that promotes hunger and weight gain, while those in good company enter your bloodstream more slowly, stabilizing your metabolism.

Let's examine what happens when you eat any type of sugar from carbohydrates, be they from white flour or phytonutrient-rich wild blueberries. When you eat sugar of any kind, your pancreas produces a master metabolism hormone called *insulin*. Insulin's job is to help sugar get into your cells. Once sugar is in the cells, it can be turned into energy by your mitochondria (the energy-burning factories in your cells). So insulin is designed to help you use the sugar you eat, or, if you eat more than you need, store it for later use.

At its best, the interaction between your insulin level and the sugar in your blood is a finely tuned machine. You eat some sugar, and your body produces just enough insulin to metabolize it. Later you eat a little more sugar, and the same thing happens again. It is a smooth, harmonious cycle that the healthy body carries out every day without your slightest awareness.

However, problems can occur when there is too much sugar in your diet. When you regularly eat a lot of sugar, especially sugars that are quickly absorbed, the insulin levels in your blood become elevated. Over time, you can become resistant to the effects of insulin and thus need more and more of it to do the same job. This *insulin resistance* has some very serious health implications as well as a direct impact on your appetite.

Insulin resistance is very much like a drug addiction. When you are addicted to a drug, you develop a tolerance to it and hence need more and more of it to produce the same effect. When you consistently have a high level of insulin in your blood, you develop a tolerance to it. As a consequence your body's tissues no longer respond normally to the hormone. Hence your pancreas produces more of it, elevating your insulin levels even more in your body's attempt to overcome this resistance.

This turns into a vicious cycle very quickly. When you have more insulin in your blood than you do sugar, your body tells you to eat some sugar to even out the balance. But every time you eat the sugar you cause your insulin levels to go up even more, causing you to want more sugar, and on and on the cycle goes. In the meantime you are storing all the excess sugar as fat, slowing down your metabolism, and promoting heart disease, dementia, and cancer.

This is a condition known as prediabetes. It is also called metabolic syndrome, insulin resistance, and syndrome X. All are names for the same thing. In chapter 9 you will learn more about this condition and how it has a direct effect on your ability to lose weight.

For now you need to be aware that the types of carbohydrates you eat have a direct impact on how quickly and to what degree you develop insulin resistance.

Different types of carbs turn into sugar at different rates in your body. When your body is flooded with sugar, your insulin levels spike. This is a bad thing. It drives you down the road to insulin resistance much more quickly.

But certain types of carbs burn more slowly. It takes them longer to turn into sugar in your body; hence your insulin levels stay more consistent over time. This is a much healthier way to eat, and it keeps your metabolism running optimally.

So the question you want to have answered now is a simple one: Which carbs should I eat, and which ones should I stay away from? The problem is that the answer is not that simple. After all, it's not as though the choice is between sitting down at night and eating either a plate of bread or a pile of broccoli for dinner. There are carbs you'll want to stay away from and ones that you will want to incorporate into your diet—concepts we'll get into a little later in the chapter. But the key to balancing your meals so that you don't develop insulin problems revolves around another new concept: the glycemic load of a meal.

A New Key to Understanding Carbohydrates: The Glycemic Load

Over the years many different terms have been used to describe carbohydrates. The terms have changed so much that most consumers and even most doctors are confused by them. You've probably heard many of them: simple carbs, complex carbs, starches, sugars, the glycemic index . . . the list seems to go on and on. However, there is only one meaningful definition that has emerged from all the research: the *glycemic load*.

Looking at the glycemic load (GL) of the food you eat is a new way of thinking of food, meal composition, and carbohydrates that is practical and helpful for weight loss—and the best part is that it is a single, simple concept for you to understand. (The good news is that most low-GL foods are also high in phytonutrients.) Glycemic load measures the real response of your blood sugar (and hence your insulin level) to an entire meal.

The glycemic load is the effect a total meal has on your blood sugar and is not only related to the original form of the carbohydrate. In fact, if you put three tablespoons of psyllium (Metamucil) into a cola, you could change it from a high-GL to a low-GL drink! The reason is that there are many factors that determine how quickly carbs turn into sugar in your body. Not only the carbohydrates you choose for a given meal but everything else you eat in that meal (protein, fat, and fiber) affects how quickly or slowly you absorb the sugars in food.

The glycemic load is the best measure of this. It takes into account all the factors, including the effect that mixing carbohydrates, fats, protein, and fiber has on your metabolism.

Eating meals that have a high glycemic load means that the combination of foods that you eat will cause all the carbohydrates in the meal to be absorbed very rapidly and raise your blood sugar just as rapidly. On the other hand, a low-glycemic-load meal contains a combination of foods that either don't have many carbs to begin with or whose carbs are absorbed slowly and don't lead to the rapidly rising and high blood sugar levels that promote obesity and aging.

Your common sense should work well when you consider the glycemic load of a meal. Nonetheless, let's look at a few examples to help clarify how you can judge whether or not you are eating a meal that has a high glycemic load or one that has a low glycemic load.

Consider the typical pasta dinner: spaghetti with tomato sauce, garlic bread, and an iceberg lettuce side salad. As delicious as this meal may sound, it is really not that good for you if you want to lose weight or stay healthy. This meal is heavy in carbs that turn into sugar in your body immediately.

Eating white bread along with pasta is like adding a tablespoon of sugar to a cola. The proportion of protein and fat is minimal, as is the side salad (think of the normal portion sizes at your local hometown Italian restaurant). There is nothing in this meal that is going to slow the rate at which these carbs become sugar.

Now let's consider another meal that is equally high in carbohydrates but has a low glycemic load: chili beans with a side of fresh steamed vegetables drizzled with olive oil and balsamic vinegar. Beans of any kind are actually very high in carbs but have a very low glycemic load because of all the fiber.

Add to this a side of fresh vegetables, and you have a meal that has more carbs than you might expect.

However, this meal has a low glycemic load. The reason? Beans contain a tremendous amount of fiber, which slows down the rate at which carbohydrates are absorbed. This has enormous health implications. If you eat *this* meal, you are still consuming a high carbohydrate load. But these carbs burn long and slow inside your body, allowing it to harmoniously generate insulin and digest these sugars at a healthy slower rate without triggering the metabolic signals that promote hunger and weight gain.

If you had a choice between these two meals, you might instinctually know that you should choose the beans and vegetables and not the pasta and bread. That's why I say that judging glycemic load is largely common sense. Unfortunately, we don't always make the best choices. Once we get into the cycle of eating high-GL meals, our appetite spins out of control, taking us down the road to obesity, heart disease, dementia, diabetes, cancer, and early death.

But you don't always choose the foods that you know will help you. Why? Perhaps you are caught in the same metabolic fat trap as the Pima Indians, who, once they were introduced to eating white flour and sugar, couldn't stop eating them.

THE NUTRITIONAL SECRETS OF ULTRAMETABOLISM: LOW GL AND HIGH PI

The foundations of the nutritional principles and the food recommended in the UltraMetabolism Prescription are two simple ideas that can guide all your food choices. These are the glycemic load (GL) and the phytonutrient index (PI). Eat foods with a low GL and a high PI, and you will ensure a healthy metabolism and optimal health.

Low-GL/High-PI Foods	High-GL/Low-PI Foods
Vegetables	Flour or flour products
Fruits	Refined grains (white rice)
Beans	Sugar in any form
Nuts	Processed foods
Seeds	Junk food
Olive oil	Large starchy potatoes
Whole grains	
Teas	
Herbs and spices	

Note: Animal products, including animal protein (meat, chicken, fish, pork, lamb) and fats (butter, lard, etc.), fall into a unique category. They have no phytonutrients because by definition they are not from plants. So they have a low PI, but they *are* absorbed slowly and don't raise the blood sugar quickly or very high. But if you eat only animal fat and protein, you will miss out on the most important and powerful compounds for healing and healthy metabolism known to science: *phytonutrients.*

The Pima Indians: A Tale of Two Carbs

The Pima Indians, who live in Arizona, had a metabolism that was exquisitely adapted to their environment. They had evolved to thrive perfectly on particular foods that exist only in a desert environment. One hundred years ago they were thin and fit and suffered none of the diseases of Western civilization such as obesity, heart disease, or diabetes. Yet in a single generation they became one of the most obese populations in the world, second only to the Samoans. Eighty percent of them have adult onset, or type 2, diabetes by the time they reach 30, and they have a life expectancy of only 46 years. What happened? Did they suddenly mutate and get the obesity gene?

No. The answer is much more complicated than that. Traditionally the Pima diet consisted of whole grains, squash, melons, legumes, beans, and chilies supplemented by gathered foods including mesquite, acorns, cacti, chia, herbs, and fish. It was a diet of whole, unrefined, and unprocessed foods. Interestingly, this diet is very high in carbohydrates. Yet the Pima were a fit and healthy people until their diet changed.

Over the course of one generation, they went from eating this traditional diet to eating the "white menace." Both diets were equally high in carbs, so what changed so radically that a people that was once fit and fabulously attuned to its environment became one of the most obese in the world? Was it a switch from a higher-fat, higher-protein Atkins- or South Beach–type diet to a diet high in carbs that led them to gain enormous amounts of weight? Absolutely not. The traditional Pima diet, the diet that had kept them thin and healthy, was very high in carbohydrates.

Scientists estimate that the traditional Pima diet, although seasonally variable, was approximately 70 to 80 percent carbohydrate, 8 to 12 percent fat, and 12 to 18 percent protein.[2] But before you go back on your old bagel, potato, pasta, and low-fat muffin diet, let's consider what kind of carbohydrates they ate.

The traditional cuisine the Pimas ate was filled with meals that had a low glycemic load. The carbs the Pimas were used to eating converted into sugar in their bodies relatively slowly. These carbs included tannins (a group of astringent and bitter compounds found in the seeds and skins of grapes that slow oxidation and aging), vitamins, and minerals. They were dense with phytonutrients, nutrients, and antioxidants. In essence, these carbs were "good" carbs. They sent the Pimas' bodies positive signals of weight loss and health.

But when their diet changed, the *information* contained in the food they ate changed dramatically. They went from eating food that sent messages to balance their weight to eating food that directed their bodies to gain weight and develop diabetes. These new carbs had a high glycemic load, digested much too quickly, and caused their insulin levels to spike; all these are signs of "bad" carbs.

As a consequence of this change in diet, they went from being a healthy, amazingly well-attuned people to one of the fattest, least healthy peoples on the planet in a single generation. The Pimas didn't have much of a choice about this because their traditional ways of life and their diet were taken away from them. You do.

What a High-Glycemic-Load Diet Has Cost the American People

The comparison between your situation and that of the Pimas may not seem particularly relevant to you. After all, you weren't raised eating chia and mesquite. But if you look at what eating a high-glycemic-load diet has cost the American people, the comparison might be brought into perspective for you.

As mentioned in the last chapter, our obesity rate has *tripled* since 1960. Interestingly, this date corresponds directly to two major dietary shifts in our culture. As discussed in the last chapter, this was right about the time the concept of a low-fat diet being good started being promoted by the government, food corporations, and the pharmaceutical industry. As unfounded as that shift was, it had a major impact on the health of the American people. For the first time since the industrial revolution, the life expectancy of the average American is declining despite all our medical and public health advances. This is a very sobering thought and directly related to the increasing rates of obesity. Obesity will take nine years off the life of the average person.[3]

When the amount of fat in our diet was reduced so dramatically, what do you think it was replaced with? If you guessed high-glycemic-load carbs, you are right. This was the second major shift in the American diet that happened around this time. The absence of fat meant that we had to fill that slot

in our diets with something. That something was highly processed carbohydrates that are cheap to produce and hence very profitable.

This whole issue was reinforced and our intake of these bad carbs soared
in the 1990s, when the U.S. government published the original food pyramid, which recommended that high-glycemic-load carbs in the form of
bread, rice, and cereal should be the single biggest component of your diet.

The consequence? Today two thirds of our population is overweight and
obesity will soon overtake smoking as the single greatest cause of death in
the country.

Was Dr. Atkins Right?

At this point in the book what I have told you is that you should eat more
fat and less bad carbs. So the question you may be left with is: Was Dr. Atkins
right? Is a low-carb, high-fat diet where it's at? Unfortunately, that is the
wrong question.

It all depends on the *type* of fat you eat and the *type* of carbohydrates you
eat. If you eat bacon, cream, and steak, you will not get the same weight loss
benefits as you would if you ate other, healthier fats, such as olive oil, nuts,
and fish. *And* you may create a whole new set of problems, including an increased risk of cardiovascular disease and stress on your blood vessels, bones,
and kidneys, not to mention constipation, bad breath, and hemorrhoids.

If you eat high-glycemic-load foods and meals made up of cinnamon
buns, bagels, and soda, your metabolism will be quite different than if you
eat whole, unrefined carbohydrates such as vegetables, beans, nuts, seeds,
whole grains, and fruit.

So Dr. Atkins wasn't quite right. But the answer to the question illustrates an important point that we need to discuss more fully before we close
the chapter. There is one simple principle that will help you effortlessly eat
food that has a low glycemic load and is high in phytonutrients—the perfect prescription for a healthy metabolism, for quick and sustained weight
loss.

The Importance of the Whole-Food Diet: The Secret to Understanding Carbs

The best part about this chapter is that you don't have to memorize it. All
you have to do is remember this simple rule: carbs should come from
whole, unprocessed plant foods.

Whole foods come in a million varieties—high fat, low-fat, high-carb, low-carb, high-glycemic-index, low-glycemic-index, complex carbs—and they are *all* good. (Note: Not all whole foods are carbs.) The key is to eat whole, real, unprocessed food, as close to its natural state as possible. If you've been eating mountains of highly processed foods such as candies and crackers and decide to make a switch to whole, real unprocessed foods such as vegetables, fruits, whole grains, beans, nuts, seeds, olive oil, organic, range, or grass-fed animal products (poultry, lamb, beef, pork, eggs), and wild, smaller fish such as salmon, starting right now, you will lose weight.

These foods contain an abundance of obesity-fighting chemicals, vitamins, and minerals that will accelerate your metabolism and plenty of fiber that will slow the absorption of sugar into your bloodstream.

One critical key to the glycemic load is *fiber*. Fiber slows the absorption of sugar into the bloodstream? What is fiber, anyway?

Fiber is the secret key to eating meals that have a low glycemic load. Let's examine it further.

The Fiber Factor: The Secret of the Glycemic Load

As we have been discussing throughout this chapter, the whole issue of whether a low-fat or low-carb diet is better or whether you should eat Atkins, Ornish, the Zone, or South Beach misses the point entirely. The secret factor behind the glycemic load that hardly anyone talks about is *fiber*. Some studies have shown that low-carb diets result in more weight loss than low-fat diets, while some studies show exactly the opposite. How can that be? The missing link is fiber, or what used to be called roughage and was thought to have no value in the human diet.

What we now know is that fiber is a powerful substance that has the ability to help you lose weight; lower your blood sugar and cholesterol; reduce your risk of cancer, heart disease, and diabetes; and reduce inflammation (a topic we will address fully in chapter 11).

Fiber is like a sponge that soaks up fat and sugar in your gut and both slows and prevents some of their absorption. Since your body has to work harder to digest it, it slows down the digestive processes. Think about the difference between eating an apple and drinking apple juice. They contain the same basic nutrients, but the apple, in its whole, high-fiber form, requires more breakdown time and metabolic effort. That means the whole apple has a lower glycemic load than the apple juice. A high-fiber diet lowers the glycemic load of any meal by slowing the rate at which the sugar you are eating is digested, thus improving your metabolism.

Fiber is one of the main factors that determine the glycemic load of a meal and the effect a meal has on your waistline. Remember the earlier example where we compared the beans to the pasta? The reason the beans have a lower glycemic load is that they are packed with fiber. It takes your body much longer to digest beans full of fiber than to digest pasta full of quickly digested sugars.

Can You Eat a Low-Fat, High-Carb Diet and Still Lose Weight?

One study recently discredited the low-carb craze. The study compared a low-fat diet to a low-carb diet. The low-fat diet group did better in every way. But the author never really played up what was hidden in the fine print: he gave the low-fat group a special supplemental fiber shake a few times a day, which provided them with more than 60 grams of fiber. The average American consumes 8 to 12 grams per day, and the American Heart Association recommends 25 grams a day. The weight loss in that study had nothing to do with the amount of fat or carbohydrates; it was the fiber that made all the difference.

Why? *Because it lowered the glycemic load of the diet.*

The fiber, not the low-fat diet, was the secret of that study. In another study, Dr. Ludwig again showed that people who ate more fiber lost more weight and lowered their insulin and cholesterol levels and clotting factors, all of which promote heart disease.[4] The fiber intake was more significant than the total fat intake.

What Does All This Add Up To?

With the decline of the Atkins empire, many Americans have realized that man cannot live on meat alone. Every major food producer has created a line of "low-carb" foods. Unfortunately, carbohydrates are the most important dietary substance we consume.

The low-carb fad promotes a myth that is harmful. It is not supported by scientific evidence and represents yet another case of our culture promoting ideas that aren't founded in reality or even common sense.

The good news is that you don't have to worry about fat and carbs anymore. You can eat both a high-fat and a high-carbohydrate diet if you want, or any combination thereof, as long as it has both a low glycemic load and a high phytonutrient index. In fact, you can forget just about everything in this chapter as long as you choose whole, unprocessed foods that are full of fiber, antioxidants, vitamins, minerals, phytonutrients, and healthy fats. The UltraMetabolism Prescription is founded on these foods. And if you choose to eat them instead of the highly processed, nutrient-depleted foods consumed by most Americans, I guarantee you will see results.

Summary

- ⁘ Carbohydrates are the single most important food you can eat for long-term health.
- ⁘ Low-carb diets are no more effective in causing weight loss than low-fat diets are.
- ⁘ Most good carbs come from whole plant food. The key to eating good carbs is eating whole, unprocessed food.
- ⁘ These plant foods are filled with important phytonutrients that can't be replaced by any other food.
- ⁘ All the terminology related to the low-carb craze is outdated. There is only one concept you really need to focus on: the glycemic load of your meals.
- ⁘ Stick to carbs that have a low glycemic load, and you will feel healthier and lose weight faster.
- ⁘ Eating food that is quickly turned to sugar by the body is stressful and increases adrenaline and cortisol, which causes you to gain weight.
- ⁘ Eating food that quickly turns to sugar also makes you eat more and gain weight because it produces more insulin, which signals your brain to eat more.
- ⁘ The secret of choosing the best carbs to eat is to choose many whole, unprocessed foods.
- ⁘ Unprocessed foods have much more fiber in them than do processed carbs.
- ⁘ Fiber is the secret key to a diet with a low glycemic load.

CHAPTER 5

THE SUMO WRESTLER MYTH:

Skipping Meals Helps You Lose Weight

Becoming a Sumo Wrestler

*M*ichael *was a musician. At 44 he discovered the guitar and devoted his life to practicing, playing, and living a musician's life in Hawaii—staying up late, playing in clubs. He would wake about noon or so, full from a late meal the night before. Meals were his reward for playing for hours.*

By the time he ate he was really hungry, so he ate large quantities. The foods were generally healthy, so he didn't think there was any problem. After he ate, he went to bed. During the day he didn't eat much, just a snack here or there.

By 49 years of age, he had put on a good 30 pounds and was more fatigued during the day than he used to be. His triglycerides skyrocketed, and his good cholesterol (HDL) plummeted. He had always exercised and loved to ride his bike up and down the hills in Hawaii. He was fit but fat.

Getting him to eat breakfast, eat before he played, and not eat three hours before bed allowed him to lose the extra 30 pounds without even trying. It also normalized his cholesterol results, and immediately stopped all his sugar cravings. He found out that when *you eat is just as important as* what *you eat!*

Is a Sumo Wrestler Born or Made?

By now you know that this culture has many misconceptions about carbs and fats, and you know that starving yourself won't help you lose weight. But there are other reasons that obesity is such an epidemic in this country. One of the reasons that Americans are getting to be as big as sumo wrestlers is that we actually eat like sumo wrestlers.

Have you ever wondered how people get to be as big as sumo wrestlers? Think about it: How do the Japanese, whose physique is relatively small, produce these behemoth specimens? Are they genetic mutants bred to gain weight? Or are they a product of a deliberate, ancient method of creating huge warriors?

Actually, most sumo wrestlers start out as small boys from the countryside. It's their lifestyle and the way they eat that makes them into the mammoth men they are. Let's examine the average day of a sumo wrestler and see how we in the United States follow a very similar way of life and end up with a very similar result.

A Day in the Life of a Sumo Wrestler

Sumo wrestlers wake early, usually about five in the morning, skip breakfast, and engage in five hours of a strenuous exercise session called *keiko*. Then they bathe and eat their main meal of the day, a protein-rich stew called *chanko-nabe*. Chanko-nabe is a healthy chunky soup made of seaweed, bonito flakes, cabbage, leeks, shiitake mushrooms, bean sprouts, tofu, noodles, chicken, pork, salmon, scallops, eggs, rice, and soy sauce. They eat this with even more rice and then top it off with a few beers or sake.

After this huge meal, they spend the next several hours napping. Then they wake and soon are ready for dinner and bed again. The next day, they wake and do the same thing all over again. They do this every day for years on end.

This regimen of exercise, eating, and sleep leads to enormous weight gain, particularly when carried on for long periods of time. So what's the biology behind it? Why does this pattern cause such dramatic weight gain that sumo wrestlers reach a staggering 400 to 700 pounds?

The Biology Behind the Sumo Diet

There are some specific ways the sumo diet causes ordinary people to gain extraordinary amounts of weight. The first is that sumo wrestlers never eat breakfast. They wake up, and the first thing they do is start exercising vigorously.

This combination of skipping breakfast and then training really hard for five hours means that by the time they get to eat, they are starving. As a consequence, they overeat—they eat much more food than they need to feel full.

To optimize our health and weight loss, we need to eat breakfast, to spread our food intake evenly throughout the day, and not to eat for at least two hours before bed. A recent study[1] found that almost 3,000 people who lost an average of 70 pounds and kept it off for six years ate breakfast regularly. Only 4 percent of people who *never* ate breakfast kept the weight off.

The only difference between the two groups was that the group who lost weight ate breakfast and the other group did not. They both consumed the same number of calories and the same types of food. It turns out that it's not only the type of calories you consume that is important in losing weight and maintaining weight loss, but the time of day you eat as well.[2]

Another important point is that sumo wrestlers go to sleep almost immediately after they eat. Sleeping immediately after you eat is a sure way to pile on the pounds.

When we are sleeping, we are in a healing and repair mode, a storage and growth mode. Those of you who have teenagers know that they wake up taller in the morning than when they went to sleep. This is because they produce more growth hormone while they are asleep. You produce more growth hormone when you sleep, too. The problem is that once we finish growing up, we grow out!

The body slows down its metabolism during sleep, and any undigested food left in your system is stored for later use. How is it stored for later use? Usually as fat.

If you can avoid eating for two to three hours before bedtime, you give your body the time it needs to digest the food. That way, when you go to sleep you are not storing the calories you just consumed as fat. Instead, you burn them to keep your bodily functions active while awake.

America: A Country on the Sumo Diet

Does skipping breakfast and eating a large meal just before sleep sound familiar? It should. It's the American way. We consume most of our daily calories shortly before bed. We rarely eat breakfast. We hardly make time to eat during the day, and by the time we get home we are literally starving, we often consume more than we need and then go to bed or sit in front of the television or computer while munching on more snacks. Then we do the worst thing possible, the thing that is guaranteed to make us gain weight: we go to sleep. No wonder we are looking more and more like sumo wrestlers every day.

What's worse is that the foods we eat aren't nearly as healthy as the diet sumo wrestlers stick to. We have an unhealthy eating pattern coupled with unhealthy foods, and we expect to remain thin, fit, and healthy. It simply isn't realistic.

What's the take-home lesson here? There are a few.

The first is that you need to spread your eating evenly throughout your day. Also, make sure you eat breakfast every day, and don't eat anything for two or three hours before bed (unless you want to become a sumo wrestler). I will discuss this more in chapter 9. For now, suffice it to say you need to eat breakfast. When you eat breakfast you're "breaking your fast," just as the word suggests. It is an important way to tell your body that sleep time is over and it is time to get your metabolism into gear.

In addition, it is important to not to go to sleep immediately after you eat. Eat earlier, or wait for a couple of hours after you eat before you go to bed. If you don't, you're just asking your body to store the calories you just consumed as fat.

Summary

- ⁘ Eat breakfast daily.
- ⁘ Spread your food intake and calories throughout the day.
- ⁘ Wait at least two to three hours after your last meal before going to sleep.

CHAPTER 6

THE FRENCH PARADOX MYTH:

The French Are Thin Because They Drink Wine and Eat Butter

The French Paradox Meets the American Paradox

*J*oel was 25 and had ballooned to more than 300 pounds. I asked if he had ever lost weight or tried to lose weight in his life. He said, "Only once, and I didn't try." I asked him what he had done, and he recounted the story of his year in France. He had lived with a student chef who had gone to the market and cooked delicious meals of fresh, real food every day.

They ate, drank wine, and slowly enjoyed the course of their meals every day. Joel didn't have a car at the time and didn't want one, because he loved walking from place to place and seeing France. He wasn't on a "diet," but he lost 35 pounds that year.

Life is the art of drawing sufficient conclusions from insufficient premises.

—SAMUEL BUTLER

So if the sumo diet doesn't work, what about the French diet? The French have a reputation as a people that knows about food, what to do with it, and how to eat healthfully. After all, everyone knows that the French eat more fat, drink more wine, and yet suffer less heart disease and are less obese than Americans, right?

Well, that's only part of the story. The truth is that in the 1960s the French diet received 25 percent of its calories from fat. That's not even close to the amount Americans consumed in fat during the same time period (you will remember that Americans in the 1960s got 42 percent of their calories from fat).

It's true that today the French consume more fat than Americans (the French now eat a diet that consists of 40 percent fat while the American diet is 34 percent fat), but while they are seeing increased rates of diabetes, heart disease, and obesity as the fat in their diet increases, they still don't come close America's problems with these issues.

How can we make sense of these confusing statistics? It's true that in the past the French seemed to eat a diet high in fat, yet were healthier than Americans. While that may be changing now, the French and Mediterranean diets and lifestyles still contain valuable lessons that can help us stop the epidemic of overeating and obesity.

There are a few, very important distinguishing characteristics between the French or Mediterranean and American diets. While wine has received much credit, it doesn't tell the whole story either. Small amounts of antioxidant-rich red wine may be healthful, but a bit too much is harmful, so it isn't drinking wine that makes the French healthier.

The real story is this: the French eat *real* food, they eat *less* food, they eat food more *slowly* than Americans do, and they *walk* more than we do. Let's see how each of these factors makes it easier for them to keep the weight off.

Real Food

You may wonder exactly what I mean by "real" food. What is real food, anyway? We Americans don't eat imaginary food, that's for sure. So what makes the French diet more "real" than ours?

The traditional French diet has always been built around food that is fresh, full of nutrients, and minimally processed. It is a whole-food diet. Traditionally the French shopped daily in the local market for fresh vegetables, fruits, fish, meats, and dairy products. They ate food grown by local farmers in soil that was naturally nutrient-rich, not enriched by man-made products and chemically fertilized, and not imported by truck, boat, or plane across thousands of miles.

Historically the French didn't eat snack foods, junk foods, or fast foods. They ate healthy monounsaturated oils from olive oil and nuts and omega-3 fats from fish, as well as smaller amounts of saturated oils from cheese. They still eat many more vegetables than the average American, and these are rich in phytonutrients. They eat more beans and fresh fruit than we do. And they don't have to worry about the ingredients on the labels of the foods they eat, because they don't eat many foods that require labels.

A few generations ago, the food culture in the United States was more like this. My grandmother's nutritional advice was simple and powerful: "Buy fresh, eat fresh." I have extended that same sentiment to a saying of my own: "If it has a label, don't eat it." After all, how many labels do you see growing in nature? When was the last time you bought a peach with a bar code on it?

All the Food Your Grandmother Ate Was Organic, Whole Food!

Actually, what my grandmother knew about food was fairly profound. In an earlier age, people in this country ate a whole-food, organic, farm-raised, grass-fed, pesticide-, hormone-, and antibiotic-free, local and freshly picked diet just as the French did, and they were healthier for it. There was a consciousness about eating fresh foods grown locally.

All of our grandparents probably ate whole, *real* food. Forty percent of Americans lived on farms in 1900, compared to the 2 percent who live on farms today.[1] All of the food was organic, all the chickens free-range, all of the beef grass-fed, and none of the food was genetically engineered. Take a road trip to Anywhere, U.S.A., and try to find real food today. It is a huge challenge to be in a town, on a highway, or anyplace else in America and find real food to eat. You will have lots of choices for garbage foods: McDonald's, Burger King, KFC, Wendy's, Denny's, Taco Bell, Dunkin' Donuts, and more, but you will be hard put to find a restaurant that will serve you fresh vegetables or fish, or free-range chicken.

Today when you take a trip, you have to take real food with you. It's not your fault that you can't find real food in the average American town. The food industry conspires to keep real food off the shelves. Why? It's simply not as profitable to sell vegetables, fruits, and nuts as it is to sell Twinkies, candy bars, and chips.

There are now 320,000 processed foods and beverages competing for our food dollars in this country; 116,000 of them have been introduced since 1990. Most of these are candy, gum, and snack foods. Thirty billion dollars is spent marketing these foods to both adults and children. There are more than a quarter-million fast-food restaurants. More money is spent every year on fast food than higher education, new cars, and computers combined! If those resources were spent promoting a real, whole-food diet and making it possible for people in this country to find food that works with rather than against their metabolisms, we would not be facing the health crisis that we are today.

Remember, we evolved eating real, whole food, not the fake foods that have become common fare—foods that are commercially produced, packaged, processed, chemically altered, and full of nonfood ingredients, including hydrogenated oils, colorings, preservatives, and fillers. No amount of promotion by the food industry can change this basic truth.

Less Food

The French also eat *less* food.[2] They eat smaller portions. Beginning in the 1970s, there have been major changes in U.S. agricultural policy, especially subsidization of corn production and farming. This has led to an overproduction of corn and an extra 500 calories available every day for every person in America, mostly in the form of *high-fructose corn syrup*. We now have 3,800 calories available to each of us every day, almost double what the average woman needs to maintain her weight. Many of you might remember the time when a soft drink came in 6- and 8-ounce sizes. Now the average size is 20 ounces. Of course, the bottle clearly states that it is two and a half servings. But who shares?

A recent study[3] examined the average portion sizes consumed of specific food items (salty snacks, desserts, soft drinks, fruit drinks, french fries, hamburgers, cheeseburgers, pizza, and Mexican food) in relation to their eating location (home, restaurant, or fast food). Between 1977 and 1996, food portions increased significantly for meals eaten both inside and outside the home. The energy intake and portion size of salty snacks increased by 93 calories (from 1.0 to 1.6 ounces); of soft drinks by 49 calories (13.1 to 19.9 fluid ounces); of hamburgers by 97 calories (5.7 to 7.0 ounces); of french fries by 68 calories (3.1 to 3.6 ounces); and of Mexican food by 133 calories (6.3 to 8.0 ounces).

We eat more because food producers add "value" by supersizing everything without regard to the consequences for our health or our waistline. The French, on the other hand, are accustomed to smaller portions. They don't deprive themselves, but they stay away from the gargantuan portion sizes Americans have become accustomed to. This holds true for foods made at home, items purchased in the supermarket, and portions scooped up at the few "all-you-can-eat" restaurants in France.

Slow Food

Another thing the French do is eat *slow* food; that is, they eat food more slowly. And they eat food that is absorbed more slowly—the real, whole, fresh foods mentioned earlier.[4]

Many Americans eat quickly, and we often eat when we are doing something else. Food is less associated with pleasure and often simply considered fuel for the body. We eat unconsciously without many of the social and environmental cues that help us relax, digest, and metabolize our food.

Studies observing French people eating at McDonald's found that it took them longer to eat french fries or a Big Mac than Americans eating the same food. So the French both eat *less* and eat *more slowly* than Americans. But if they continue to eat Big Macs, they will catch up with us in health and weight problems. A Big Mac has 600 calories—almost half of the average woman's resting metabolic rate—and half of those calories come from unhealthy trans and saturated fats (33 grams). One Big Mac contains half our daily allowance of salt and has 50 grams of high-glycemic-load carbohydrates in the form of a white-flour bun.

The Metabolic Power of Pleasure:
How Eating Slow Food Speeds Up Your Metabolism

Food has traditionally been associated with family, friends, nourishment, celebration, and pleasure. Many Americans have lost these connections. People in this country have started to judge food based simply on what feels pleasurable to them at the moment. They have abandoned the many other important things that food and the eating experience can contribute to their life, such as talking with friends, communing with their families, or telling stories around the dinner table. Focusing on what brings immediate gratification is often done to compensate for the lack of pleasure people feel in their jobs, daily living habits, relationships, or social status. This separation of pleasure from the eating experience contributes to problems with the metabolism of food.

What we have lost, and what the French maintain, are the social aspects of the meal: eating with family and friends in a relaxed atmosphere. In other words, the pleasure that you experience when you eat food is also important to your metabolism. We need to be in a relaxed state for the nervous system of our gut or digestive system to work properly.[5] Eating while we are stressed out makes us fat, both because we don't digest our food properly and because stress hormones slow metabolism and promote fat storage, especially of belly fat.[6] We also tend to overeat when we eat quickly, because it takes the stomach twenty minutes to signal the brain that we are full.

A pioneering nutritional psychologist, Marc David, author of *Nourishing Wisdom* and *The Slow Down Diet,* reported a comical story about eating food fast while eating fast food. One particular man he treated was intent on losing weight. The patient had only one condition: he would not give up the two giant fast-food hamburgers he ate every day for lunch. He was a busy man and didn't have time to prepare anything else. In fact, he purchased his supersized burgers at the drive-through section and would scarf them down before he left the parking lot.

Rather than fight with the man about the need to stop eating fast food, Marc David simply recommended that he eat his fast food slowly and enjoy every bite. He suggested sitting inside the fast-food chain, relaxing and smelling his superburgers, chewing, tasting, and enjoying every bite. He asked him to eat consciously and with pleasure. The man complied and came back the next month looking significantly thinner. What had happened? When he had finally slowed down to taste and enjoy his food, he had realized, "This tastes absolutely disgusting! I'll never eat at another fast-food restaurant again."

Walking

When was the last time you walked to the store to buy your groceries? One of the charms of a French town is that you can walk everywhere. Most were built long before the automobile and were designed on a human scale where everything is close enough to walk to. In fact, the roads in a great many European towns are too small for cars. In contrast, American suburbs and living communities were generally created for cars, not people, so we have never acquired the habit of strolling.

Walking to the store to buy your food or going for a stroll after you have eaten means that you are exercising more, burning off the calories you consume. This is a great way to stay fit, and one that America has divorced itself from in the creation of urban sprawl.

There are some important lessons Americans can learn from the French about food. But they are *not* to eat more saturated fat and drink more alcohol. We need simply to eat real food; eat smaller portions; eat more slowly; eat with pleasure; and walk after dinner. If you do so, you will feel more satisfied, less hungry, and better nourished, *and* you will lose weight and improve your metabolism.

Summary

- The French are thinner and healthier because they eat whole *real* food, not commercially processed foods containing trans or hydrogenated fats and high-fructose corn syrup.
- Eating food quickly or while you are stressed out can make you fat, especially in your belly area.
- Try some of the strategies the French use: eat real food, eat less food, and eat more slowly to give your metabolism a helping hand.
- Create pleasure around mealtimes and spend more time savoring

food with family and friends. Pleasurable meals actually speed up your metabolism.

❖ Go out and take a stroll or walk to your grocery store. If you can't do this, find little ways to increase your activity: take the stairs, not the elevator or escalator; park as far away from where you are going as possible; give up your remote control; take a walk after dinner.

CHAPTER 7

THE PROTECTOR MYTH:

Government Policies and Food Industry Regulations Protect Our Health

Food Politics

*A*lice had three children between the ages of 8 and 12, each with his or her own after-school schedule—violin lessons, art class, soccer, Irish step dancing—and she found she spent more time in her minivan than her house, and certainly more time there than her kitchen. She had become a "drive-by eater" (as had her children). They ate meals together in the minivan often and only rarely around the dinner table. Even when they did eat at the table, it was often pizza or take-out subs. Thank heaven, she said, that there were fast-food restaurants at every corner! She was not alone; 30 percent of Americans eat at least one meal in their car every week, and one in five breakfasts is a McDonald's breakfast!

Slowly, then a little more quickly, the pounds started coming on. Her children also started gaining weight. The abundance and convenience of poor-quality food lacking in nutrients but high in sugar calories and trans fats was not worth the trade-off. But, with her busy schedule she didn't know any alternative. In fact, there is hardly anywhere to find quick and convenient food that isn't harmful.

But as we worked together, she learned how to shop and quickly prepare real, whole-food snacks and meals that she could take with her and feed her children with. And I gently suggested that she wouldn't be a bad mother if her kids didn't do every activity, that there was value in staying home as a family, preparing meals, and slowing down. Junk in, junk out, I told her, and she listened. Not completely, and not right away; but she found that with a little planning she could start eating real food every day and was able to lose her minivan pounds.

It is difficult to think of any major industry that might benefit if people ate less food; certainly not the agriculture, food product, grocery, restaurant, diet, or drug industries. All flourish when people eat more, and all employ armies of lobbyists to discourage governments from doing anything to inhibit overeating.

—M. NESTLE[1]

At the beginning of the twentieth century, Americans cooked and ate most of their meals at home. By the year 2000, they ate more than half of their meals outside the home or ate processed food they didn't cook at home. This is not an accident or an unintentional result of busy lives or the breakdown of the family unit. It has been an intentional transformation of the nutritional landscape by an industry spending phenomenal resources to market inexpensive food of poor nutritional quality to more and more people, regardless of the effect on their health or waistlines. I live in a small rural town in Massachusetts, and there are five McDonald's within ten miles of my house. How can we win the battle of the bulge with more than 13,000 McDonald's in the United States alone?

How Industry and Government Sabotage Our Health

The food industry generates more than a trillion dollars in annual sales, accounts for 12 percent of the U.S. gross national product (GNP), and employs 17 percent of the country's labor force. It spends more than $33 billion a year on marketing alone. Seventy percent of those dollars go to pushing fast food, convenience foods, candy, snacks, soft drinks, alcoholic beverages, and dessert. Only 2.2 percent of those marketing dollars are spent on advertising for fruit, vegetables, grains, or beans.[2]

Meanwhile, the U.S. Department of Agriculture (the USDA, the government agency responsible for agriculture and food policy) spends only $300 million on nutrition education, which goes mostly to research or agricultural extension projects that reach few people. And oddly enough, the same government that can't find money to fund public health campaigns to promote the scientific principles of good nutrition could increase agricultural subsidies from $18 billion in 1996 to $28 billion in 2000.[3]

Why is the USDA, which represents the agriculture industry—one of the biggest food industries in this nation—responsible for food policy? The USDA committee responsible for setting our dietary recommendations is populated by many "experts" who work for the food industry, hardly an objective scientific body. The Department of Health and Human Services (HHS) *should* be the guardian of our health. The current system is like putting the drug companies in charge of the Food and Drug Administration (FDA) and letting them set drug policies and approve new drugs. It would be unthinkable. Yet that is the way it is with our current food and food policy guidelines.

The food industry is pushing us face forward into the overconsumption of foods that trigger weight gain, obesity, and all the known diseases of aging, including heart disease, diabetes, stroke, cancer, and Alzheimer's disease. These conditions are *not* the inevitable result of aging. They are related to the quality of our diet.

This idea may seem strange: food can make us sick. And it is one that is foreign to the policy makers who are in charge of setting food guidelines in this country. They ask: If food is just energy, why should it matter whether that energy comes from a cheeseburger or grilled tofu? You know the answer to this question—why don't they?

The recent "cheeseburger bill" passed in Congress prohibits lawsuits against the food industry for making us fat or sick. Why create a law to protect the food industry when the government is entrusted to protect citizens? A recent study[4] of the causes of death in the United States found that poor diet and physical inactivity contribute to approximately 400,000 deaths annually, a one third increase from 1990 and almost as many deaths as those caused by smoking. Obesity will soon overtake smoking as the leading cause of all deaths in the United States. Don't you think this recent rise in eating-related health disorders is due at least in part to the advertising dollars major food companies put into convincing us to eat low-quality food items? Apparently the U.S. Congress doesn't.

Consider this: the largest food companies in the world are also tobacco companies. Two examples are RJR Nabisco and Altria Group (which owns Philip Morris and Kraft Foods). (In 1999 RJR Nabisco split to avoid tobacco boycotts, which also affected their food sales. In 2000 Philip Morris bought Nabisco.) It makes one wonder if it will take the U.S. population and its government as long to realize that the food these companies are feeding us is destroying our health as it did for us to become aware that the cigarettes they were selling us were killing us. It's a strange coincidence that obesity is now overtaking smoking as the leading cause of death in this country when it is the same companies foisting the products on us that caused these problems in the first place. The tobacco companies and the junk-food companies are the *same* companies!

The Origins of Our Diet: Real Food Matches Our Genes, Fake Food Does Not

While it is true that we were designed to thrive on food, what we are eating now is not food. The food industry is promoting "food" that does *not* turn on genes that promote health and optimal metabolism. Loren Cordain, Ph.D., of the Department of Health and Exercise Physiology at Colorado State University, in his remarkable article "Origins and Evolution of the Western Diet: Health implications for the 21st century,"[5] outlines just how far we have come from the diet we were evolved to eat and to which our genes are well adapted. This diet has changed dramatically in the ten thousand years since the agricultural revolution, but our genes have changed little.

Our original diet had seven characteristics, none of which applies to our modern diet:

1. A low glycemic load (or low sugar and refined carbohydrates)
2. More omega-3 fats from wild food such as fish, wild game, and wild plants
3. A balance of protein, fat, and slowly absorbed carbohydrates
4. An abundance of vitamins and minerals
5. Many alkaline foods (plants), which prevent our blood from becoming too acidic
6. A low level of sodium (salt)
7. A high fiber content

Everything about our industry-driven diet in the twenty-first century works against our genes and promotes not only obesity but every age-related and chronic disease that afflicts us. We would do well to reacquaint ourselves with the diet to which our bodies are best adapted, so we can live in harmony with our genes (making it much easier to fit into our jeans). This is what our bodies thrive on, the diet of our hunting and gathering ancestors, not the diet of drive-through fast-food windows and the candy shelves of the local 7-Eleven.

Obviously you are not likely to start foraging for roots and berries or hunting for your dinner, but there are very practical ways to change your diet to meet your genetic needs. Choose real, whole, unprocessed food, such as an abundance of fruits and vegetables, beans, nuts, seeds, and whole grains; eat small wild fish such as wild salmon, sardines, and herring; avoid food with added salt; avoid fake foods, particularly those containing high-fructose corn syrup and hydrogenated fats. This will bring you as close as possible to our evolutionary wild diet.

Toxic Sugars and Toxic Fats

The food we eat now is chemically altered to promote shelf life and increase consumption, not good health. There are two major ingredients in most of the prepared foods you eat: corn and soy. They're vegetables, so what could be bad about that? The problem is that corn and soy are transformed in the laboratory into toxic foods, unknown to human biology, when they are turned into the supersugars and superfats known as *high-fructose corn syrup* and *hydrogenated soybean oil.* Read the labels of the food on your shelves, and you will likely find these culprits among the list of ingredients.

These two man-made chemicals are responsible for the dangers of almost all the fast and processed foods in this country. They are associated with "empty calories"—calories that don't contain vitamins, minerals, antioxidants, phytonutrients, fiber, or essential fats. That's right, high-fructose corn syrup and hydrogenated oil have absolutely *no* nutritional value, and they are found in almost every processed or packaged food you consume these days.

Trans fats are inert. They never go rancid as butter or regular vegetable oil does. Crackers or cookies made from them can stay on the supermarket shelf for months, even years, which is great for the profits of their creators but absolutely destructive to your health. If you forgot a can of shortening on a shelf in your pantry for ten years, it would still be the same as the day you bought it. Would you eat an apple that sat on your shelf that long without changing? Now just imagine what those toxic substances are doing inside your body!

Toxic sugars and toxic fats are insidious and ever present. And they are particularly bad for your metabolism and weight. They make you gain more weight and cause you to eat more and feel hungrier than their unprocessed counterparts do.

High-fructose corn syrup is used to sweeten almost everything these days, including soft drinks. It wasn't even in our food supply before the 1970s. Since 1997, we have increased our consumption of soft drinks from 23.3 gallons to 54 gallons per person each year. Before the mid-1970s, sodas were sweetened with sugar; now they are sweetened with high-fructose corn syrup. These supersugars quickly enter your bloodstream and trigger hormonal and chemical changes that make you feel even hungrier. And as we learned in chapter 4, insulin surges start a cascade that tells your brain to eat more and your fat cells to store more fat.

Hydrogenated oil is used to preserve everything from cookies to crackers to salad dressings. It is in almost every package of food that you pick up on your supermarket shelf, because it is the primary agent that allows the food to be stored on a shelf in the first place. It is a universal danger and incredibly hazardous to your health. These trans fats bind to a spot in your cells that blocks your metabolism, slows fat burning, increases cholesterol, and leads to insulin resistance or trouble balancing your blood sugar. This leads not only to weight gain but to every other major health concern.

These issues don't seem to concern the food industry. They continue to use these deadly products in every food item they create despite the scientific evidence that tells us they are deadly.

Weapons of Mass Expansion: The War on Whole Foods and Why the Food Pyramid Is Killing Us

How can Americans find food that will nourish them, promote optimal health, and support a well-functioning metabolism in this vast nutritional wasteland? How can they find good food when our government spends tens of billions of dollars to subsidize the production of toxic foods and the food industry spends more than $30 billion annually to develop and market them? To make matters worse, the Department of Health and Human Services spends only $300 million a year on public health education (trivial amounts of which are spent on nutritional education). We are up against "weapons of mass expansion," and we are losing the battle.

The unfortunate truth is that government policy is closely tied to industry. The ill-conceived 1992 Food Pyramid developed by the USDA, the same organization whose mission is to promote and support agriculture, was based on weak science and commercial interests. It advocated the consumption of six to eleven servings of bread, rice, cereal, and pasta and told us to eat fat sparingly. We have now learned that this is the prescription for insulin resistance (prediabetes). (See chapters 3, 4, and 9 for more information on why this is so.)

The scientists who developed it were affiliated with the food industry, as opposed to independent university or medical researchers, who might have had a more objective point of view. This Food Pyramid became the basis of the school lunch program, hospital food programs, nursing home menus, and all federally funded institutions that serve food. Our government told us to eat a diet that serves business, not health, a diet that makes us sick.

This was followed up by a revised set of food guidelines introduced in 2005, but these are only a mild improvement. Rather than educating us about the dangers of refined carbohydrates and sugars, they meekly advise us to "choose carbohydrates wisely." The new food pyramid, while an improvement, is still controlled by industry influence.

On the new government Web site www.mypyramid.gov, I entered my personal information. I was told to drink three glasses of milk a day. While I will grant that this is better than recommending three sodas a day, milk contains saturated fat and creates acidity; in addition, for 75 percent of the world's population, consumption of dairy products leads to digestive problems such as irritable bowel syndrome. Why, then, would the government standard for prescribing food guidelines to people recommend that I drink three glasses a day? It simply doesn't make any sense.

Dr. Walter Willett, in his book *Eat, Drink, and Be Healthy,* which is based on more than twenty years of nutritional research at the Harvard School of Public Health, takes issue with the Food Pyramid and turns it upside down, placing healthy fats at the bottom and encouraging us to eat mostly vegetables, fruits, whole grains, legumes, nuts, fish, eggs, lean poultry, and only a little meat, sugar, bread, and dairy. But after the government-sponsored Food Pyramid was introduced, we reduced our intake of fat, increased our consumption of bread, rice, pasta, and cereals, and doubled our rates of obesity.

So What Can I Do?

Clearly, government food policy needs to change. Dr. Marion Nestle, professor and former chair of the Department of Nutrition, Food Studies, and Public Health at New York University, sums up the changes that are necessary in our food and nutrition policy quite well: "Existing food policies could be tweaked to improve the environment of food choice through small taxes on junk foods and soft drinks (to raise funds for anti-obesity campaigns); restrictions on food marketing to children, especially in schools and on television; calorie labels on fast foods; and changes in farm subsidies to promote the consumption of fruits and vegetables. The politics of obesity demand that we revisit campaign contribution laws and advocate for a government agency—independent of industry—with clear responsibility for matters pertaining to food, nutrition and health."

While these changes would dramatically improve the food landscape in this country, they are likely still a way off. You need ways to change what you eat right now, so that you can start losing weight and feeling better today.

UltraMetabolism and its 7 Keys to the new science of healthy weight will teach you how to make the best eating choices and lifestyle decisions despite the misinformation provided by the government and the food industry. It will empower you with the information you need to change your health and your life right now.

Summary

+ The food industry in this country spends enormous amounts of resources creating foods that aren't healthy and promoting these foods for your consumption.
+ Government policy doesn't generally support good nutritional advice any more than the food industry does.
+ The reason government policy doesn't provide good nutritional

advice to the public is that it is closely tied to food industry interests.

❖ The government Food Pyramid that advised us to eat six to eleven servings of bread, rice, and cereal a day has significantly contributed to the epidemic of obesity.

❖ Poor diet is the second leading cause of death and will soon overtake smoking as the number one cause of death in America.

❖ You should turn the Food Pyramid upside down by putting healthy fats (omega-3 fats from fish and flaxseeds and monounsaturated fats from olive oil) at the bottom, and eat mostly vegetables, fruits, whole grains, legumes, nuts, fish, eggs, and lean poultry and only a little meat, sugar, refined carbohydrates, and dairy products.

❖ Concentrating on a diet based on whole, real, unprocessed food is the best way to avoid the health risks associated with eating fake foods that damage your health.

PART II

The 7 Keys to the New Science of Weight Loss

The Owner's Manual for Your Body

✣

To know truly is to know by causes.

—FRANCES BACON

The 7 Keys to Successful Weight Loss

There are 7 Keys to weight loss, and all of them work together to open the door to vitality, good health, and successful long-term weight loss. None requires taking drugs.

1. **The first key** is to control your appetite and metabolism by understanding how the brain, gut, and fat cells communicate with one another through hormones and brain messenger chemicals called neuropeptides to drive your eating behavior.
2. **The second key** is to understand how stress makes you fat and how to overcome its effects.
3. **The third key** is to control inflammation, a hidden force behind weight gain and disease.
4. **The fourth key** is to prevent cellular "rust," which interferes with metabolism and causes inflammation.

5. **The fifth key** is to learn how to turbocharge your metabolic engine to turn calories into energy more efficiently.
6. **The sixth key** is to make sure your thyroid, the master metabolism hormone, is working optimally.
7. **The seventh key** is to detoxify your liver so it will properly metabolize sugars and fats and eliminate toxins and toxic weight.

Understanding these 7 Keys is critical to long-term, sustainable weight loss that doesn't depend on deprivation or punishment but is based on nourishing and caring for your body. While one weight loss guru or another might emphasize any one of these elements, these seven factors have never before been integrated into a complete program.

This is the first comprehensive medical and clinical perspective to outline and provide a program that addresses *all* of the most important aspects of metabolism and weight loss. Some of these concepts, such as inflammation and oxidative stress, have never before been identified as keys to weight loss. UltraMetabolism provides a view of the entire landscape of how to achieve and maintain a healthy weight.

The way we live and the unconscious choices we make every day sabotage our best intentions. Armed with the latest information science has to offer, we can take back our health and find that fit person inside waiting to come out. Working with our bodies instead of against them, we can learn to stop acting in a way that triggers a flood of molecules sweeping us down a slippery road of survival behavior designed to make us store fat. That is the promise of UltraMetabolism.

CHAPTER 8

CREATING ULTRAMETABOLISM IN YOUR BODY:

An Overview of the 7 Keys

Fixing All the Keys to Create UltraMetabolism

*L*auren came into my office complaining of fatigue, weight gain, and chronic digestive problems. She was a 54-year-old married consultant with no children who traveled about 220 days out of the year. She described herself as overstressed and overworked. At five feet six inches tall, she weighed 265 pounds and looked very much like an apple, with skinny arms and legs and a big round middle. She had gained most of the weight after early menopause at age 43.

She led a hectic lifestyle, running around all day to meetings and propping herself up with coffee and sugar. Her typical breakfast consisted of a banana and coffee. She craved carbohydrates like bread and pasta, rice and potatoes. And when her day finally quieted down around eight or nine o'clock at night, she was starving. After eating a large dinner, she would experience severe reflux or heartburn and would take an antacid such as Zantac just to be able to sleep without her food coming back up. Her busy, unbalanced lifestyle and subsequent fatigue did not allow for quality, let alone the minimum quantity, of exercise. Every morning she woke up exhausted, and she frequently fell asleep on airplanes and in front of the television.

While Lauren was at Canyon Ranch, I found she had a number of significant, undiagnosed problems that accounted for her symptoms of weight gain, fatigue, and digestive problems. I identified a bacterial stomach infection, Helicobacter pylori, that can cause heartburn, gastric reflux, and even generalized inflammation. Inflammation is one of the keys to UltraMetabolism. It is often a major unexplained factor causing weight gain or preventing weight loss. In fact, her level of inflammation, measured by a special blood test called C-reactive protein, was very high at 5.3 milligrams per deciliter (normal is less than 1).

Her "good" (HDL) cholesterol was very low, and her total cholesterol to HDL ratio, which normally should be less than 3.0, was 8.35. She was severely insulin-resistant, a state induced by high levels of refined carbohydrates in the diet, which caused her to produce huge quantities of insulin, which made her even hungrier for starchy carbohydrates. She had also developed a fatty liver, common in people with *blood sugar problems, which meant she had built up even more toxic weight because she couldn't detoxify properly. Impaired detoxification, or trouble with getting rid of our own or environmental toxins, is another major key leading to weight gain. Her metabolism, tested during exercise, was 40 percent slower than it should have been; in other words, she burned 40 percent less calories than she should have when exercising.*

I prescribed an eating plan that switched Margaret's metabolism from storing fat to burning it. She went on the UltraMetabolism program: eating protein for breakfast, eliminating the white menace (white sugar, white flour, and refined carbohydrates) to reduce her blood sugar swings, eating more omega-3 fats from wild fish and flaxseeds, and performing moderate exercise in the mornings. We treated her stomach infection with antibiotics and had her take multivitamins, fish oil, and lipoic acid (a special nutrient that improves metabolism) to help balance her blood sugar, as well as an herb, milk thistle, to heal her fatty liver.

Lauren returned three months later, 53 pounds lighter, three sizes smaller on top, and one size smaller on the bottom. Her blood sugar and insulin levels had improved dramatically. Her total cholesterol had dropped like a stone, and her good cholesterol had increased. Her liver and inflammation tests had returned to normal. She now wakes up full of energy and has no more heartburn. Lauren has her life back again.

Giving Your Genes the Right Instructions

Lauren's story is not an unusual one. Over the course of the last twenty years, I have seen many patients like Lauren, people who come to my office having tried everything to lose weight. They did not know that most of what they were doing was so steeped in the mythology that surrounds health and weight loss in this culture that they had little hope of changing their lives using outdated and scientifically unfounded methods. They go on the UltraMetabolism Prescription, they lose weight, they are measurably healthier, and they feel better about their lives.

For the first time, their bodies are given the right messages, the right information, and the right instructions for their genes. How do you know what instructions to give your body?

An Instruction Booklet for Your Body?

Don't you wish you had come with a little instruction book, perhaps strapped to your leg as you emerged from the birth canal? Wouldn't it be wonderful if it contained all the directions for what you should eat, how much you need to exercise, whether you require more protein or fat, when you should absolutely never eat sugar, or why you might need twice as much vitamin C as your next-door neighbor?

The science of medicine is advancing very rapidly, particularly in this new era of genomics, and soon we will be able to provide the answers to most of these questions. As the science advances, we will be able to prescribe diets, lifestyles, drugs, and nutrients that can help you maximize your genetic potential and minimize your risk of disease and obesity.

This section will explore some of the new vistas in medicine as they apply to weight regulation and obesity. These are the 7 Keys needed to open the door to the new science of weight loss. Each of them is important, and all interact; however, in some individuals, one or more of these principles predominate.

Understanding them will help you succeed in managing not only your weight but also your long-term health and well-being. They are the keys to reversing disease, being set free of chronic symptoms, and creating optimal health. Not only will you experience effortless weight loss and create a healthy metabolism, you will also address the underlying causes of chronic disease. The consequence of these changes will mean a new feeling of life and energy.

Let's unlock the door to understanding why it's not your fault if you're overweight or fat. The menus and recipes in part III are nutritionist- and chef-designed to make this practical, doable and delicious. You will have the opportunity to discover new foods and enjoy familiar ones in new ways that have many healing properties.

Everyone needs to discover his or her *own* instruction manual. The best way to do this is to use the quizzes and steps in part II to guide you in personalizing UltraMetabolism for yourself.

Finding Your Own Key

Understanding all of the 7 Keys is necessary to understand your metabolism. They are essential to help you create UltraMetabolism. At times, you will wonder how all this complex science can be made practical. But that is the wonder of the body.

Making subtle changes in your lifestyle, meal composition, meal timing, and the quality of the food you eat, exercising moderately, taking supplements that support your metabolism, taking saunas, and using stress management can have a powerful and dramatic effect on your weight and your overall health with very little effort. These principles will help you stop the seesaw of weight gain and loss that comes from an imbalanced metabolism. When you are in balance, the body takes care of itself.

The sections toward the beginning of each key will guide you with questionnaires and tests to delve more deeply into specific areas that may be more problematic for you as an individual. After assessing your own particular situation, you can follow the suggestions for correcting these problems or choose to get professional help.

You may have a hormonal or medical reason for being overweight, and all the dieting or exercising in the world won't fix it until you find the real cause. You may have an undiagnosed thyroid problem, or you may have a common hormonal imbalance found in women that causes weight gain, facial hair, irregular periods, and infertility called polycystic ovarian syndrome (PCOS).

Such problems need to be properly assessed, diagnosed, and treated to get your metabolism working again. The questions and suggestions provided at the end of each key will teach you how to get your metabolism back in balance. Information about medical testing for problems with each of the keys is included in case you need to explore the cause of your problems further. More information about choosing the right supplements, plus resources needed to implement all the suggestions in this section, can be found in the appendix.

Important Note on the Quizzes for the 7 Keys

During each visit with a patient, I review an extensive list of questions that helps me identify key areas of imbalance or metabolic problems. I have created self-scoring questionnaires that will help you identify where *you* have a problem. This is critical to helping you customize UltraMetabolism and personalize your dietary, lifestyle, supplement and testing recommendations. Please refer to this general interpretation guide, which applies to the questionnaires in each of the keys of UltraMetabolism.

Interpretation

Here is how to interpret your score and what to do about it.

Low: 0 to 3

❖ Do the basic UltraMetabolism Prescription.

Moderate: 4 to 6

✢ Follow the steps in each chapter to overcome the metabolic problems associated with that key and optimize the UltraMetabolism Prescription for your needs.

High: 7 and above

✢ Do the UltraMetabolism Prescription; customize it using the specific recommendations in this chapter. If you score high (7 and above) you would benefit from taking the additional tests noted in the last step. In addition to further testing, I strongly recommend that you seek professional medical assistance. To find a practitioner familiar with the principles outlined in *UltraMetabolism* go to www. ultrametabolism.com.

How to Use the Suggestions in Each Key

The UltraMetabolism Prescription is the basis of getting the right instructions to your genes. However, we are not all the same, and many people have undiagnosed medical problems. Such problems can be identified in each of the keys. Maybe you have trouble with appetite control, have a low-functioning thyroid, or are very stressed. Using the quizzes and the steps in each key will help you personalize the program and overcome these problems.

If you find you have problems in more than one key (and for each key you can score low, moderate, or high), you can combine suggestions from different chapters. Because the body is a unified system, similar suggestions are found in multiple chapters: to eat whole foods, exercise, relax, and take similar supplements. This is actually good news. The body is designed to have all systems function when given specific health-supporting inputs—the right food, movement, nutrients, and so on.

Therefore it will be easier than you think to combine the approaches to problems identified in different chapters because many of the recommendations are the same—you won't have to eat twice as many whole foods, relax twice as much, or double-dose on supplements and herbs.

Use the chapters in part II as a resource guide, a way of navigating through difficulties. As best I can, I have given you the experience of my years in practice helping people who have problems that are difficult to solve.

A Message on Special Foods, Herbs, and Supplements

You can safely include the foods recommended in each of the keys, and you can try some of the extra herbs and supplements specified (in addition to the basic recommended supplements I detail in part III). The higher you score in a key, the more likely you are to benefit from taking the additional herbs and supplements suggested for that problem. Experiment. Explore. Consider working with an experienced nutritionist or physician who can guide you further. Details about each of the supplements and herbs recommended are provided in appendix B. Even more detailed information about foods, supplements, extra recipes, specialized testing, finding a practitioner, and other online tools are available at www.ultrametabolism.com.

In chapter 16 I provide the basic principles of a lifelong healthy way of eating. It is a clear set of guidelines for foods to include and foods to avoid or limit to create a healthy metabolism, lifelong health, and well-being. This is your new road map for healthy eating.

Taking Additional Tests and Seeking Professional Help

Finally, keep in mind that in some cases the health problems you face may be aided by taking some additional tests. You may also need to seek professional medical help. If you score high on any of the quizzes in part II, you will see that I recommend you do just this.

This doesn't necessarily mean that you are having a health emergency. But it does mean that you should very seriously consider seeing a doctor who can help you test for the conditions I discuss in this book and help you implement the program in it.

The reason many people continue to struggle with weight loss is that their recommendations have not been personalized. There is no one perfect diet, supplement, pill, or exercise program for everyone. That is the exciting promise of the latest scientific research on our genes, *nutrigenomics*.

We are at the dawn of an era of personalized medicine in which physicians will one day be able to take a drop of your blood, put it on a microchip, and, based on your genes, guide you in choosing the best nutritional program, supplements, and lifestyle program for *you*. We are not quite there yet, but after working with thousands of patients I have found that by doing some intelligent detective work using the tests I outline in this book, we can find the root or cause of their difficulties—whether it be a chronic illness or trouble losing weight. The tests are often a critical part of finding the answer to your chronic struggle with weight.

I have offered detailed information on www.ultrametabolism.com/tests about where you can get additional tests (in some cases you can order the tests for yourself online), learn about the best laboratories to do the tests, and find a physician qualified to help you with your problems.

Unfortunately, most physicians, due to no fault of their own, have no training in nutrition or in treating obesity, even though it is the most prevalent disease in our society (perhaps that's because it wasn't considered a disease until 2002). The information in this book is on the cutting edge of scientific research about weight loss, so in the typical HMO setting you may not be able to find a doctor who can help you.

Keep in mind that nobody knows your body better than you. When looking for a doctor, bring the information you have found in this book with you and talk with the doctor about it.

You can go ahead and start the UltraMetabolism Prescription now and customize whatever parts of it you need to, based on the results of the quizzes that follow. Then, when you go to your doctor, you can tell him or her about the research you have been doing about your health and your weight. If he or she is unwilling to consider what you are saying, you might consider looking for a different health practitioner.

Create Your Own UltraMetabolism

The information that follows will help you succeed in managing not only your weight but also your long-term health and well-being. These are the keys to reversing disease and being set free of the chronic, painful, and annoying symptoms we live with every day such as fatigue, headaches, stress, digestive problems, joint pains, and postnasal drip. They are the keys to creating optimal health.

Not only will you experience effortless weight loss and create a healthy metabolism by following this program, you will also address the underlying causes of chronic disease. The consequence of these changes will be a new feeling of life and energy.

The problem with our current approach to weight loss and health problems is that everyone is treated equally and given one prescription for weight loss: "Eat less, exercise more." As you know, that isn't the whole story.

We are all genetically and biochemically unique. UltraMetabolism will help you identify the hidden keys to *your* metabolism and turn on the genes that will make you healthy and fit. Using this approach, you can create a personalized program on a foundation of scientifically based, sound nutritional principles.

You will also be given the tools to navigate further into your own story using quizzes and suggestions for special testing. The result will be the creation of an UltraMetabolism—*yours!*

Summary

+ Part II summarizes the critical keys that underlie nearly all weight problems (and most chronic diseases).
+ Using the quizzes in each chapter, you will learn if you need to focus on correcting problems in that key.
+ Following the steps in each chapter in part II for foods, supplements, herbs, and tests, you will help correct those problems.
+ If you need further help, suggestions for testing and further help are offered.

A SPECIAL NOTE FROM DR. HYMAN

While I was writing this book, I realized two things. First, since I am constantly refining my methods based on what I learn from treating my patients, I need a way to get that updated information to you. A good example of that is the list of recommended resources where you can find high-quality supplement manufacturers whom you can trust. Second, I realized that it might be helpful to readers if some of the information were made easier for you to print out and take with you wherever you need it. A good example of that is the shopping lists for the recipes, which you'll want to take to the store with you.

The solution that I came up with is to provide this information in a freshly updated, downloadable companion guide that you can access from www.ultrametabolism.com/guide. After you download that guide, you'll have:

+ A handy resource that will include updated quizzes so you can continue to assess your health
+ Shopping lists, so you can easily print them and bring them to the store with you
+ A list of the kitchen supplies that you'll need to purchase
+ A list of the foods that you'll want to avoid and foods you'll want to enjoy on your quest to attain UltraMetabolism
+ Additional information on the tests and laboratories that I recommend so you can hand your doctor the information he or she

needs to properly assess your health and identify what's been holding back your weight loss efforts

❖ A food journal that will help you track your weight loss efforts, which is something that I've found has really helped my patients tackle any problems that might arise as they are getting started on UltraMetabolism

❖ A list of the resources that I've included throughout this book so you always have the latest information right at your fingertips

Science is constantly producing new breakthroughs that can be used in our efforts to fight our bulging waistlines. This guide will allow me to pass along to you very quickly anything from the latest research that I've proven to work with my patients. This companion guide is free and can be downloaded by going to www. ultrametabolism.com/guide. You should download this guide now, as you might want to take notes in some of the spaces provided rather than writing in this book.

CHAPTER 9

CONTROL YOUR APPETITE:

Harnessing Your Brain Chemistry for Weight Loss

The Power of the Gut-Brain-Fat Cell Connection

I was invited to the Time/ABC Summit on Obesity in 2004, a gathering of the greatest minds in science, industry, government, and the media, and even all the weight loss gurus: Dean Ornish, Barry Sears, Andrew Weil, Arthur Agatston, and a representative of Atkins, Inc. The surgeon general, the secretary of health and human services, Peter Jennings, and other notables were present.

We discussed science, theory, politics, and strategy for three days, but nothing had a greater impression on me than the taxi ride from the airport, which I shared with two others, one a lawyer and lobbyist representing the food industry, the other a weight loss doctor who had once weighed more than 400 pounds.

Of course I couldn't help but comment on how the food landscape of the average American town provided a toxic eating environment—fast foods and convenience stores at every turn. Even if someone wanted one, it would be difficult to find a healthful snack or meal. I complained that the 64-ounce supersize sodas and re-fined starches that were so accessible simply fueled food cravings and the obesity epidemic.

The lawyer for the food industry disagreed. It is all a matter of personal choice, she said. You can just say no. She was thin. The quiet and mildly overweight doctor in his midfifties in the front seat finally jumped in. He recounted his own long struggle with weight.

During his residency he had lived on coffee and sugar to enable him to work the grueling hours typical of medical education. He had ballooned to more than three hundred pounds. He experimented with every diet imaginable—Atkins, Jenny Craig, Slim-Fast, NutriSystem, protein shakes for a year, HCG (placental hormone) shots, and more. He had finally resorted to gastric bypass when he yo-yoed back up past 400 pounds.

After the surgery, he quickly lost about 80 pounds but then figured out that even though he couldn't eat a lot at once, he could satisfy his cravings by eating M&M's one after the other. He even failed gastric bypass: he had found a way to undermine it by eating M&M's one at a time until he gained back every pound that he had lost after the operation.

He described the feeling of an insatiable hunger and craving always gnawing at him. Then one day came fen-phen, one of the first pharmaceutical appetite suppressants. He tried it. Suddenly, he said, it was as if a switch had been turned off in his brain. He wasn't hungry anymore. It was at that moment he realized that his drive to eat was hormonal or chemical in origin.

But as he found out, drugs are not a solution to weight loss. When the dangers of fen-phen came to light (it was eventually shown to cause heart disease), he stopped, but he learned to use food as his medicine, rather than his poison. He was able to control his appetite, lose weight, and keep it off without the drugs by controlling his appetite through careful use of meal timing and composition. Though I never treated him as a patient, it is one of the most remarkable examples of the power of the gut–brain–fat cell connection. Getting the signals mixed will lead to misery. Getting them right will lead to good health and weight loss.

Appetites Out of Control

Have you ever wondered how your body controls your weight? How your body knows when to eat and when to stop eating? What happens inside the body to tell you that you are hungry or full?

The answer to each of these questions is your appetite control system: a complex set of chemical interactions among your brain, nervous system, metabolic hormones, special fat cells, and immune system. I call this the gut–brain–fat cell connection.

These chemical interchanges tell you whether or not you need food and compel you to eat. When they are working properly, they are an elegant machine that pinpoints when you need energy and asks you to consume calories to obtain that energy. But when they go haywire (and there are many ways for them to go haywire in our current food climate), they cause you to eat when you don't need to, which not only leads to weight gain but contributes to almost every other health problem we face.

Appetite control is as tightly regulated as your heartbeat or breathing. Imagine the effects of eating just 100 extra calories a day over the course of a year. Every 3,500 calories is equal to one pound of fat. Over the course of a year the average person eats 900,000 calories. If you exceeded your needs by just 2 percent, or 18,000 calories, you would gain five pounds by the end of the year. The average American gains about 20 pounds between the ages of 25 and 55. This is due to an excess of only 0.3 percent of calories per year over thirty years. This incredibly small difference leads to significant weight gain over time.

One of the major reasons average Americans gain this extra weight is that their appetite control system is out of balance. The chemical interactions among the various systems in the body that tell them when they are hungry have been disrupted. Rebalancing and then fine-tuning that system is what this chapter is about. In it I will show you how your appetite control system works and what you can do to repair and optimize it. Doing this is the first step to creating UltraMetabolism in your body.

But before we dive into your appetite, take the following quiz to see if you are currently having problems regulating the amount you eat.

In the rest of this chapter you will learn how your appetite control sys-

---

HOW GOOD IS YOUR GUT-BRAIN-FAT CELL COMMUNICATION?

Score 1 point each time you answer "yes" to the following questions by putting a check mark in the box on the right. See page 78 for how to interpret your score.

	YES
Have I gained weight around my belly?	☐
Do I crave sugar or carbohydrates?	☐
Do I feel tired after eating a meal?	☐
Do I eat fewer than five servings of fruits and vegetables a day?	☐
Do I eat fewer than 30 grams of fiber a day (the average American diet has about 8 grams) from beans, nuts, seeds, vegetables, and fruit?	☐
Do I skip breakfast?	☐
Do I eat within three hours of going to bed?	☐
Do I sleep less than eight hours a night?	☐
Do I mostly eat carbohydrates alone, rather than combining them with fat and protein at every meal?	☐
Do I eat high-fructose corn syrup (found in almost all processed foods and drinks)?	☐
Do I eat fewer than three times a day?	☐
Do I feel stressed on a regular basis?	☐

If you don't want to write in this book, I've included this quiz in the companion guide that you can download by going to www.ultrametabolism.com/guide.

tem works, as well as a way you can start tuning it up to give you Ultra-Metabolism.

The Four Parts of Our Appetite Control System: Understanding the Biology of Your Appetite

Your appetite control system has four basic parts. They are:

1. **The wiring of your nervous system:** The autonomic (automatic) nervous system or wiring connecting the brain, gut, and fat cells★
2. **Weight control hormones:** Metabolic hormones, including the hormones and molecules made by your fat cells
3. **Command central messengers:** Brain messenger chemicals called neuropeptides†
4. **Molecules of inflammation:** Messenger molecules of the immune system called *cytokines,* produced in the fat cells (and white blood cells and liver cells) with wide-ranging effects

These components work together to communicate among all the organs and tissues responsible for managing your weight and keeping you alive. Their signals flow among your stomach, intestines, liver, pancreas, fat cells, endocrine (hormone) system, brain, and autonomic nervous system. Good communication equals a healthy metabolism.

The dance of molecules among these organs and tissues is what makes you eat or stop eating, and consequently what makes you gain or lose

---

★ The autonomic nervous system controls all of our automatic survival functions, such as heartbeat, breathing, temperature regulation, metabolism, and appetite. It has two different components. The first, or sympathetic, part is triggered under stressful conditions and makes you gain weight. The parasympathetic part relaxes and calms you and helps you lose weight. As we will see, this nervous system plays a large role in weight management.

† New messenger molecules are being discovered every day. They all speak to one another to control appetite. The important thing to understand is that when they are in balance your weight stays in balance, and when they are out of balance you will tip the scale. The UltraMetabolism Prescription is designed to keep them in balance. Here is a list of the most important molecules involved in appetite control: the ones made in the pancreas and the fat cells are leptin, insulin, adiponectin, visfatin, and resistin; the ones made in the brain are NPY, melanocortin, and CART (cocaine- and amphetamine-regulated transcript); the messenger molecules made in the stomach include grehlin, PYY, proglucagon products, and CCK.

weight. Even small disturbances in this system can lead to significant changes in weight over time. Because the system was designed in a time of scarce food resources, it is very efficient at making you eat more and not so efficient at making you eat less. As you saw in the example earlier in the chapter, eating only 100 extra calories a day will lead to a weight gain of five pounds in a year.

Why not just count calories, then? That would seem the sensible solution if it were simply the amount that we ate that affected our weight. But do you think you can count calories accurately? Studies of top nutritionists have shown that even they can't accurately estimate the number of calories they consume in a day. How can the rest of us be expected to count calories when the experts can't? Even if you could count calories accurately, that wouldn't solve the problem. As we have already discussed, hunger cannot be controlled by willpower alone. When your body tells you it's time to eat, you will eventually eat regardless of what your willpower says.

The key to controlling your appetite is learning how to create harmony among all the parts of your metabolism that make up your appetite regulation system.

The Music of Weight-Controlling Molecules: Creating Harmony

Molecules that communicate with our nervous system control our appetite, our food intake, and how we metabolize that food. We are learning more every day about how these molecules communicate. They have many names, they come in different forms and from different places in the body, and there are a few master metabolism hormones, but no matter how big or small, they all work together to create an intricate web that is your metabolism. These molecules are everywhere in your body and are produced by many different systems, but the main centers of activity are the brain, the gut, and the fat cells. The molecules produced by these cells are *hormones, neurotransmitters,* and *cytokines.*

Hormones are the messenger molecules of the endocrine system, including your fat cells! Neurotransmitters are the messenger molecules of the nervous system, and cytokines are the messenger molecules of the immune system, also including your fat cells! There are other systems in the body that have the ability to produce these messenger molecules as well. But these are the main places they come from.

These molecules are really all one finely tuned, hopefully harmonious, system that determines your health and metabolism. Research has just begun to uncover the story that identifies which of these molecules make you feel full, which ones make you hungry, where they come from, what makes them go up or down, and how all this controls your eating behavior. But there are some things we already know for sure.

When your stomach is empty, it secretes hormones that tell your body and brain you are hungry. Your brain then prepares the stomach to receive something good (similar to Pavlov's dogs salivating at the sound of a bell). You even begin to secrete insulin just thinking about food.

When you eat, the food enters the gut and releases yet more hormones, preparing it for digestion. As the food makes its way into your bloodstream, more messages coordinate your metabolism, telling your pancreas to produce insulin.

Your fat cells in turn send hormonal messages back to your brain to stop eating, along with signals from your stomach that you are full. Your liver then processes fat and sugar and helps coordinate storage or burning.

This entire process occurs invisibly without the slightest awareness on your part. When this process of communication is out of balance, it wreaks havoc on your system. You get hungry after you have just eaten, you store fat when you should burn it, your body starts ignoring the normal control signals for appetite and metabolism, and the result is weight gain and disease.

You can directly influence the complex melodies of your hormonal system by following these six steps:

- **Step 1:** Compose the perfect meals.
- **Step 2:** Eat early and often.
- **Step 3:** Enjoy foods that control your appetite and avoid foods that send it out of control.
- **Step 4:** Use herbs to optimize your hormonal balance.
- **Step 5:** Use supplements to control your appetite and balance the brain–fat–gut cell communication.
- **Step 6:** Consider testing to find the causes of an appetite that is out of control.

If you are having problems with this UltraMetabolism key and your appetite is controlling you, you can turn the tables and start controlling your appetite by following these steps. Doing so will customize the UltraMetabolism Prescription for your own particular needs, allowing you to turn on genes that cause you to lose weight and turn off the ones that have been making you gain weight. If you scored high on the quiz in this chapter, seek out the testing mentioned in Step 6 as well as professional medical help. Let's look at exactly how you can make this six-step process work for you.

Step 1: Compose the Perfect Meals

What you put inside your body is important. Indeed, the saying "you are what you eat" is completely true. The problem is that most people have misconceptions about what choices they should make about the type and amount of calories they consume.

We have been convinced that we should be eating a low-fat, low-carb, low-calorie diet, and none of these things is true. It's the *type* of fat, carbs, and calories we consume that's most important. Of course, quantity has something to do with it too, but if you get your metabolism in order, your body will naturally regulate the amount of calories you consume, so you can stop worrying about counting calories.

What you really need to focus on is the types of food you consume and the way you balance the various kinds of food you eat. The UltraMetabolism Prescription offers a complete meal plan that incorporates all of the information in this chapter. But understanding these principles will help you digest this information for yourself so that you can make eating the Ultra-Metabolism way a natural part of your life.

Let's start by looking at some of the principles of composing the perfect meals.

Eat Real Food

The single most important thing to keep in mind when you are building your personal menus is to include as many real, whole, unprocessed foods in your diet as possible.

I don't mean to sound like a broken record, but I simply can't emphasize enough how important it is to return to the historical roots of human nutrition. In the preparation phase of the UltraMetabolism Prescription, I am going to teach you how to get rid of all the fake food you are eating right now. Doing this will be an important step in detoxifying your body and cleaning up your diet. Not only will it make you healthier, but it will help you control your appetite and make you lose weight.

For now, start paying closer attention to the labels on your food. Better yet, stop eating foods that have labels on them. Remember: if it has a label; don't eat it. You want to be consuming foods in the closest possible form that you would find them in nature.

The following are a few examples of real foods:

- ❖ Whole fruit, not canned fruit or fruit juices
- ❖ Whole vegetables, not canned vegetables
- ❖ Wild fish, not farmed fish
- ❖ Whole grains, not processed wheat
- ❖ Grass-fed beef, not feedlot beef
- ❖ Nuts, seeds, and legumes or beans, not fried or salted

If you start doing this right now, you will feel your health improve and your energy change, you will control your appetite, and you will start losing weight.

Eat the Right Fats

As you know from chapter 3, eating the right kinds of fat is important if you want to turn on the genes that help you lose weight and turn off the ones that cause you to gain weight. This has been proven in study after study. Fats are one of the key parts of our diet that control genes, weight, and inflammation.

Fats help in a number of ways. They provide satisfying, slowly ab-

> **If it has a label, don't eat it.**

sorbed energy that makes you full faster and keeps you feeling full longer. It doesn't trigger a big surge of insulin in the body, as refined sugars or carbohydrates do. And, more important, healthy fats in your meals help to lower the overall glycemic load of the whole meal by mixing with all the rest of the food in your gut and allowing it to be absorbed more slowly, even before it hits your bloodstream.

Once in your blood, the healthy fats, including olive oil, nuts, coconut oil, and fish oil containing the omega-3 fats, help form healthy cell membrane—the outer structure of every cell that controls all the signals and messages coming from everywhere in the body (including insulin). This helps your cells communicate better. Some key fats (the omega-3) enter the cells and communicate with your DNA to turn on special genes that help you increase fat burning, improve your blood sugar control, correct insulin resistance, and reduce inflammation (which is a very important part of weight loss and health, as we will see in chapter 11).

One of the best ways to help people lose weight and regain optimal health is to give them an oil change. In fact, in a recent study of all the possible treatments for preventing heart disease, including all the new drugs, fish oil was found to be the most effective prevention.[1] Ideally we could simply eat more omega-3 fats, but unless you subsist on a diet of wild Alaskan salmon it is best to supplement with purified, metal- and pesticide-free fish oil capsules.

Balance Your Glycemic Load

You will remember from chapter 4 that carbs are not inherently bad. In fact, they are probably the single most important food in your diet. Human beings evolved eating carbs—lots of them. We just aren't built to eat the highly processed carbs that are so dominant in our diet today. Learning to compose perfect meals means knowing which carbs you should eat and which ones you should stay away from.

Choosing a diet that has a low glycemic load (GL) and is rich in phytonutrients is an important way to control both your appetite and your weight. This is true for a number of reasons. First of all, foods in this category tend to be higher in fiber. As you know, fiber slows down your digestive process. That means that you will feel fuller longer.

In addition, good carbs with a low GL and a high phytonutrient index (PI) help stabilize and balance your insulin levels. This means that you won't have cravings for carbohydrates and sugar that you don't need. In turn, you will have less initiative to eat the massive amounts of sugar that most of the people in our culture consume. This will help you heal from insulin resistance and metabolic syndrome.

One fascinating example of the power of a low-glycemic-load diet to help improve your metabolism is from Dr. David Ludwig[2] of Harvard, who has demonstrated in study after study that all calories are not created equal. In a study in the *Journal of the American Medical Association,* he demonstrated that low-fat diets slow your metabolism more than low-glycemic-load diets do. The study provided 1,500 calories a day for each group and was designed to cause a 10 percent weight loss over a few months. The low-fat group had a diet similar to that recommended by the National Cholesterol Education Program for a heart-healthy diet. The low-glycemic-load diet was *not* low in carbohydrates, just low in refined carbohydrates, such as white bread, and included whole grains. In fact, 43 percent of the calories in the low-glycemic-load diet were from carbohydrates.

The results were startling. Even though both groups lost weight, those in the low-fat-diet group had a slower metabolism at the end of the study, were hungrier, and were more resistant to further weight loss. They also had more inflammation, higher levels of triglycerides, insulin, and blood sugar, and higher blood pressure. The complex signals that regulate our weight and metabolism, therefore, are controlled by meal composition. Foods that are absorbed slowly (have a low glycemic load), that are high in fiber, such as whole foods, grains, beans, and nuts—even though they are high in carbohydrates—actually keep your metabolism running faster, burning more calories and creating a harmonious balance of metabolic signals that facilitates long-term weight loss and healthy metabolism. Eating a low-fat diet slows metabolism. Eating a low-glycemic-load diet increases metabolism.

In essence, moving from carbs that have a high GL and low PI to carbs that turn into sugar more slowly and are packed with phytonutrients is a way for you to take one of the many vicious cycles we have talked about in this book and turn it into an upward spiral of weight loss and healthy living. There are three parts to doing this:

1. Eat more fiber.
2. Avoid sugar.
3. Stay away from supersugar, high-fructose corn syrup.

Eat More Fiber

Fiber is the secret key to a low glycemic load. It's like a sponge that soaks up sugar, thus making it burn more slowly in your digestive system.

Eating fiber is advantageous for a lot of reasons. The higher the fiber content of a single food or a meal in total, the harder and longer your body has to work to digest it. This is a weight loss advantage in three ways:

1. **You burn more calories.** You burn more calories just digesting your food (a process called thermogenesis) because your body has to work longer to digest the food. This "work" burns calories. That means you are actually burning more calories while you eat.
2. **You stay full longer.** You're digesting longer, so you stay satisfied longer and ultimately eat less over the course of a day than you would if you had eaten low-fiber foods.
3. **You reduce your appetite.** Slowing down absorption means slowing down the rate at which your blood sugar rises and falls; remember, you want an even metabolic keel, not a roller coaster. Keeping your blood sugar from spiking helps you reduce the insulin spike, in turn reducing your appetite.

The aim of increasing dietary fiber is to lower your glycemic load.

Like anything else, you want to balance the fiber in your diet. I do not recommend adding wood pulp to your diet as a supplement! But increasing your fiber intake is one important way that you can lower your glycemic load and control your appetite. It will help you create a metabolic balance that allows your body to function in the most efficient manner possible.

Eat Superfibers But not all fiber is created equal, and two deserve special mention: konjac root and rye.

Konjac, *Amorphophallus konjac* K. koch, is an Asian tuber or root that is full of a viscous, water-soluble fiber called glucomannan. It is extracted from the root, dried, and made into rubbery jelly and noodles; it has been used for more than a thousand years in Japan. It is five times more powerful than psyllium, oat fiber, or guar gum in lowering cholesterol.

The magic of this fiber is something called the viscosity (or the gooeyness or thickness of the fiber). It is like a dried-up sponge that expands to ten times its size when put into water. You might not want to eat the rubbery jelly unless you like flavorless gummy substances.

The root has been studied by researchers at the University of Toronto, prepared in pills or powders that can be mixed with water and taken before meals. The results have been remarkable. This fiber blocks cholesterol absorption, reduces the production of HMG CoA reductase, the principal regulator of cholesterol in the body (drugs used to lower cholesterol, known as statins, also work to reduce HMG CoA reductase, but they often have nasty side effects that konjac root doesn't), and feeds the healthy bacteria in the gut that then produce special fats which reduce the production of cholesterol in the liver.

The fiber also makes a gel in the intestine that slows the rate of absorption of your meals and thus lowers the all-important glycemic load.[3] Used in conjunction with the healthy diet outlined in this book and exercise, konjac fiber helps reduce blood sugar and cholesterol, and facilitates weight loss. (See www.ultrametabolism.com for more information on where to find konjac root fiber.)

The other fiber worth mentioning is rye fiber.[4] Included in the Ultra-Metabolism prescription are whole grains—not whole-grain flour or whole-grain products, but whole grains. Whole grains in any form are a huge help to your diet, in no small part because they increase your fiber intake.

I am all for bread. But I have a simple rule: if you can squish it easily with your hands, you shouldn't eat it. The only way to make squishy bread is to add some kind of ground flour to it, automatically raising its glycemic load. True whole-grain bread is very dense and cannot be squished in your hands.

One type of bread available made from whole grains (not whole-grain flour) is whole-grain or kernel rye bread from Germany. Studies have

shown that rye fiber has unique properties that lower blood sugar and insulin more than other fibers such as wheat fiber. Rye foods also have special
phytonutrient compounds called lignans that are chemically similar to
human hormones. These compounds have repeatedly been associated with
lowered cancer and cardiovascular disease risk in individuals whose diets
contain high levels of such phytonutrients.

Sprouted grain products are also fine. But whole grains in general help
control blood sugar and insulin and reduce the risk of obesity, metabolic
syndrome, and heart disease. The fiber factor is an important aspect of long-
term weight loss. And fiber is easy to find. It is found in all unprocessed, unrefined, whole plant foods such as beans, whole grains, nuts, seeds, fruits, and
vegetables.

Avoid Sugar at All Costs

As you know, carbohydrates, especially those with a high glycemic load,
turn to sugar very quickly. A major problem today is that we are swimming
in sugar. The average person eats about 180 pounds of sugar a year, or about
½ pound per person per day. Remember, that is the average, which means
that some people are eating a lot more. When you eat sugar, you unconsciously trigger a vicious cycle of sugar cravings, increased insulin production, increased appetite, more sugar intake, and more insulin production,
until you are in a cycle of cravings, bingeing and crashing all day long.
Eventually this leads to insulin resistance (see pages 99–100), which is a
major contributor to weight gain and rapid aging.

Refined carbohydrates and sugars are ever present in our diet. White
bread, sugar, pasta, white rice, and white potatoes are all quickly absorbed
starches with a high glycemic load that convert to sugar in our body at a
very rapid rate.[5] Consumption of these foods has increased dramatically
since the introduction of the original food pyramid in the early 1990s,
which advised us to eat six to eleven servings of rice, bread, cereal, and pasta
a day.

Soft drinks or sugar-sweetened beverages and alcohol, consumed in excess, contribute to the problem with our excessive consumption of sugar as
well. Read food labels carefully to identify sugar in other disguises. Sugar is
sugar by any other name; watch out for:

- Corn syrup and high-fructose corn syrup
- Sucrose
- Glucose
- Maltose

- Dextrose
- Lactose
- Fructose
- White grape juice or other fruit concentrate
- Honey
- Barley malt
- Maple sugar
- Sucanat
- Natural cane sugar
- Dehydrated cane juice
- Brown sugar
- Turbinado sugar
- Invert sugar

Food processors do not have to state if there is added sugar in their products. Added sugar is defined as additional sweetener or sugar put into food other than what's naturally found in it. For example, high-fructose corn syrup is often added to fruit drinks. The manufacturer is required only to list the total grams of sugar. Be careful of hidden sugars in the following items:

- Breakfast cereals
- Salad dressings
- Luncheon meats
- Canned fruits
- Bread
- Peanut butter
- Crackers
- Soups
- Sausage
- Yogurt
- Relish
- Cheese dips
- Chewing gum
- Jellies and jams
- Frozen deserts (ice cream, sorbets, and yogurts)

You don't have to avoid these items entirely. You just want to be wary of them. Remember to read food labels carefully. If you are worried that there may be hidden sugars in certain foods, do some research or look for an alternative that you might enjoy just as much. Healthier versions of these foods without high fructose corn syrup, trans fats, or added chemicals are available at whole food stores such as Whole Foods or Wild Oats.

Stay Away from the Supersugar, High-Fructose Corn Syrup

I have already mentioned that one of the unspoken dangers of our food supply is *high-fructose corn syrup* (HFCS), an ingredient of almost every sweetened or processed food that many Americans consume every day, such as sodas, canned foods, cookies, cakes, baked goods, and frozen foods. But it bears further investigation in this chapter due to the effects it has on your appetite.

HFCS is a supersugar. Its consumption increased by more than 1,000 percent, from 0.292 kilogram per person per year to 33.4 kg per person per year, from 1970 to 1990 and now represents more than 40 percent of caloric sweeteners added to foods and beverages.[6]

Not surprisingly, the introduction of HFCS into the food supply is associated with the beginning of the obesity epidemic. Other factors may certainly play a role, including reduced levels of physical activity; increased portion sizes; eating more than half our meals in restaurants, on the go, and at fast-food restaurants; changes in the types of foods eaten (moving from a whole-food, plant-based, high-fiber diet to a sugary, trans fat–rich, low-fiber, low-vitamin and -mineral diet); and the overall "toxic food environment,"[7] with its abundance of poor-quality, nutrient-poor, sweetened, and trans fat–laden food choices. However, the effect of HFCS in soft drinks and other sweetened beverages merits serious consideration as an important cause of the obesity epidemic.

WHY THE FRUCTOSE IN HIGH-FRUCTOSE CORN SYRUP IS DIFFERENT FROM THAT IN ORDINARY SUGAR

The digestion, absorption, and metabolism of fructose differ in significant ways from those of regular sugar. Table sugar, as we know it, is a combination of glucose and fructose and is known as sucrose. Glucose is the basic sugar the body uses for energy and metabolism—one of the key building blocks of all carbohydrates and often found as part of other slowly absorbed sugars found in beans and whole grains.

Fructose is also one of those building blocks, found in nature mostly in fruit, where it is packaged along with fiber and an abundance of protective nutrients. But fructose does not stimulate insulin

(continued)

secretion and the consequent increase in *leptin,* a hormone produced by fat cells that tells your brain you are full, which reduces appetite. This isn't such a problem when you eat fructose the way nature intended it to be eaten, in the form of fruit. This is because when you eat fruit, the amount of fructose you ingest is significantly lower than in sweetened beverages, and the metabolic effects of it are different because the increased intake of fiber, vitamins, minerals, phytonutrients, and antioxidants helps slow absorption and improve metabolism.

However, when fructose is processed into high-fructose corn syrup (HFCS), it is absorbed more quickly than regular sugar and enters your cells without any help. It doesn't require the help of insulin the way glucose does. Once inside the cell, it becomes an uncontrolled source of carbon (acetyl-CoA) that is then made into cholesterol and triglycerides. Basically, that means that eating HFCS makes your cholesterol level shoot straight up and causes problems with your liver that slow down your metabolism even more. It actually produces a fatty liver (just like foie gras or pâté) and is the major cause of abnormal liver function tests in this country. Also, HFCS is probably the biggest reason for the increase in cholesterol levels we have seen in our society over the last twenty years. Try cutting out HFCS, and see how fast your triglycerides and cholesterol drop.

In fact, none of the normal controls on appetite is triggered when you eat foods or beverages containing high-fructose corn syrup. When you metabolize glucose, your brain normally gets the message that you are full. That doesn't happen as readily with HFCS. You just stay hungry and keep eating more sugar or HFCS, which continues to fuel this cycle. This has been shown to lead to increased appetite, calorie, intake, and weight gain as well as high cholesterol and high blood pressure.

Steer Clear of Artificial Sweeteners Aspartame (NutraSweet),[8] neotame, acesulfame potassium, saccharin, sucralose, and dihydrochalcones are artificial sweeteners consumed by two thirds of the adult population and are a significant component of our diets. They are in many packaged foods, artificially sweetened foods, gum, candy, sodas, drinks, and mints. Read the labels and look for those names.

Questions remain about their safety, including both short- and long-term health risks. One of the side effects we know these sweeteners have is stimulation of hunger through the cephalic, or brain-phase, insulin response. As I said before, simply thinking about sugar can cause a spike in insulin, and putting something on the sweet receptors of the tongue tells the brain that something sweet is coming and to get ready by producing hormones such as insulin.

A number of studies have shown that aspartame ingestion may actually lead to increased food and calorie intake.[9, 10] This is likely because artificial sweeteners make your body produce insulin by making it think sugar is on the way. As a consequence, your body tells you to eat more sugar to balance your insulin level. Artificial sweeteners do nothing to help in this regard. They do not act as sugar does and do not balance your insulin. As a result, you end up with excess insulin in your body, so you end up eating more food to take care of this problem. This whole pattern disrupts your appetite control system in serious ways. What's worse, it can lead to insulin resistance, which has many serious health consequences.

But the problems with artificial sweeteners don't stop there. Animal and human studies show that aspartame may disrupt brain chemistry[11] and induce neurophysiological changes (altered brain chemistry due to the fact that the sweeteners mimic normal neurotransmitters) that might increase seizure risk,[12] depression,[13] and headaches.[14, 15] More research needs to be done on the total effect of artificial sweeteners on human beings. Until that research is done, you would be safest to stay away from them.

Of 166 studies on the safety of aspartame, 74 had at least partial industry funding and 92 were independently funded. While 100 percent of the industry-funded studies conclude aspartame is safe, 92 percent of independently funded research identified aspartame as a potential cause of adverse effects.

THE EPIDEMIC OF METABOLIC SYNDROME, OR PREDIABETES

Eating too much sugar or too many high-glycemic-load carbs does more than just affect your figure. In chapter 4 we discussed a condition known as insulin resistance. When you eat too much sugar in any form, your body develops a dangerous resistance to insulin. Insulin resistance not only compounds the initial injury you are doing to your metabolism when you eat excess sugar, making it that much harder to lose weight; it has also been associated with every major disease related to aging, including cancer, dementia, heart disease, and, of course, diabetes.

In addition, elevated insulin levels lead, directly or indirectly, to other metabolic abnormalities, including hypertension, low HDL ("good") cholesterol, high

(continued)

LDL ("bad") cholesterol, high triglycerides, obesity (especially central obesity), inflammation, and abnormal blood clotting. Individuals with such symptoms are at increased risk of coronary heart disease, as well as other diseases related to plaque buildups in artery walls (such as stroke and peripheral vascular disease).

Insulin resistance is also known as *metabolic syndrome, insulin resistance syndrome, syndrome X,* and *prediabetes.* These are all names for the same condition. And it affects almost everyone who is overweight to some degree or another. With 65 percent of Americans fitting that description, that accounts for almost 193 million people!

While diet is the biggest factor in the creation of metabolic syndrome, other contributing factors include physical inactivity and genetic factors, particularly a family history of type 2 diabetes.

Because this is such an enormous health problem in this country, though it is still poorly recognized, I have included the following quiz that should give you a very good idea of whether or not you have metabolic syndrome. It will also help you assess the severity of the problem.

METABOLIC SYNDROME SELF-ASSESSMENT TEST

Score 1 point each time you answer "yes" to the following questions by putting a check mark in the box on the right. See page 78 for a reminder on interpretation of your score.

	YES
Do you have a waist-to-hip ratio greater than 0.8 if you are a woman or greater than 0.9 if you are a man? (To figure out your waist-hip ratio, measure the circumference of your waist around your belly button in inches. Take this measurement and divide it by the circumference of your hips at their widest point.)	☐
Do you crave sweets, eat them, get a temporary boost of energy and mood, and later crash?	☐
Do you have a family history of diabetes, hypoglycemia, or alcoholism?	☐
Do you get irritable, anxious, tired, or jittery or develop headaches intermittently throughout the day but temporarily feel better after meals?	☐
Do you feel shaky two to three hours after a meal?	☐
Do you eat a low-fat diet but can't seem to lose weight?	☐
If you miss a meal, do you feel cranky, irritable, weak, or tired?	☐

If you eat a carbohydrate breakfast (muffin, bagel, cereal, pancakes, etc.), do you feel as though you can't control your eating for the rest of the day? ☐

Once you start eating sweets or carbohydrates, do you feel as though you can't stop? ☐

If you eat fish or meat and vegetables, do you feel good but seem to get sleepy or feel "drugged" after eating a meal full of pasta, bread, potatoes, and dessert? ☐

Do you go for the bread basket at the restaurant? ☐

Do you get heart palpitations after eating sweets? ☐

Do you seem salt-sensitive (do you tend to retain water)? ☐

Do you get panic attacks in the afternoon if you skip breakfast? ☐

Do you often get moody, impatient, or anxious? ☐

Are your memory and concentration poor? ☐

Does eating make you calm? ☐

Do you get tired a few hours after eating? ☐

Do you get night sweats? ☐

Do you frequently get thirsty? ☐

Do you seem to get frequent infections? (For example, do you have frequent colds or poorly healing wounds?) ☐

Are you tired most of the time? ☐

Have you been diagnosed with polycystic ovarian syndrome, infertility, high blood pressure, heart disease, or adult-onset diabetes? ☐

Do you have chronic fungal infections (jock itch, vaginal yeast infections, dry scaly patches on your skin)? ☐

Optional

I recommend specific testing to diagnose metabolic syndrome. You can ask your doctor for the appropriate tests. Refer to www.ultrametabolism.com for recommendations of doctors who can help you.

If you have taken the tests and have abnormal lab results, give yourself another point for each of the following by checking the box on the right.

❖ **A low HDL level (<50 mg/dl [milligrams per deciliter] for men, <60 for women)** ☐

(continued)

- ✛ High triglycerides (>100 mg/dl) ☐
- ✛ A triglyceride/HDL ratio of greater than 4:1 ☐
- ✛ Abnormal liver function tests (AST, ALT, GGT) or fatty liver ☐
- ✛ A high serum ferritin level (>200 mg/ml [nanograms per milliliter]) ☐
- ✛ A high serum uric acid level (>7.0 mg/dl) ☐
- ✛ A low serum magnesium level (<2.0 mg/dl) ☐
- ✛ A fasting blood sugar level >90 mg/dl ☐
- ✛ A fasting insulin level >8 mIU/ml (micro International Units per milliliter) ☐
- ✛ A 1 or 2 hour post-75-gram-load sugar level >120 mg/dl or insulin >30 mIU/ml ☐

INTERPRETATION

Now see whether or not you have metabolic syndrome and how severe your condition is.

1 to 5: Doing well

- ✛ Do the basic UltraMetabolism Prescription

5 to 10: Moderate metabolic syndrome

- ✛ Overcome the metabolic syndrome by optimizing the basic Ultra-Metabolism Prescription with the protocol in this chapter

11 and up: Severe metabolic syndrome

- ✛ Have additional testing done and seek professional help. You can learn about all the tests for metabolic syndrome at www.ultrametabolism.com/tests for more information about how to obtain tests and where to seek professional assistance.

MINIMIZE THE CAUSES OF INSULIN RESISTANCE AND IMPROVE INSULIN SENSITIVITY

- ✛ Follow the guidelines for foods to enjoy and foods to avoid in part III on page 212 ("Important Message for Those Who Own a Body") that are the basis of UltraMetabolism.

❖ **If you don't want to write in this book, I've included this quiz in the companion guide that you can download by going to www.ultrametabolism. com/guide.**

Now that we have addressed all of the major issues that revolve around your meal composition, let's turn our attention to the frequency and timing of the meals you eat.

Step 2: Eat Early and Often

Not only is the composition of the meal you eat critical, but *when* you eat and *how often* you eat are essential in regulating your weight. Studies[16,17] have shown that eating regular meals throughout the day, grazing versus gorging, and not skipping meals improve your chances of losing weight *and* reduce many risk factors for heart disease, diabetes, and aging in general.

Eat Often

If you eat in a regular, rhythmic pattern, at roughly the same time every day, you will eat less, burn more fat, and lower your cholesterol and insulin levels. To achieve this, you will need to eat frequently—probably more frequently than you are accustomed to. I recommend eating three meals and two snacks a day. If that seems like a lot to you, don't worry; when you eat this often the portions of food you eat will naturally be smaller. Your metabolism will be functioning better, so you will feel less hungry and thus eat less.

In addition, if you eat regular meals you will heat up your metabolic fire and burn more calories after eating. This occurs naturally through what is known as thermogenesis (literally, "heat creation"), as a series of physiological processes is kicked into gear to digest what you've eaten and turn it into energy you can use. Eating erratically, on the go, away from home, at your desk, or whenever your hectic schedule permits will lead you down the path to constant hunger, greater food consumption, inefficient energy production, low energy level, weight gain, and obesity.

Eat Breakfast

The old proverb "Eat breakfast like a king, lunch like a prince, and supper like a pauper" now has some scientific muscle behind it. Many of us think that if we skip breakfast we will reduce our overall calorie intake for the day and lose weight. Unfortunately, the opposite is true. Not eating breakfast means you will eat more the rest of the day.

In one study of people who ate breakfast, compared to those who skipped breakfast,[18] the ones who skipped breakfast actually ate more food, had higher levels of cholesterol, and were more insulin-resistant. Over time, these all lead to more weight gain, more heart disease, more inflammation, and all the other age-related diseases.

One researcher from the University of Texas[19] evaluated the effect of circadian and diurnal rhythms (the body's internal clock) of food intake. In an analysis of seven-day diet diaries of 900 men and women, the researchers found that eating more calories earlier in the day is more satiating than eating the same number of calories later in the day and reduces overall calorie consumption. If you eat bigger meals later in the day, they won't be as satisfying and you will consequently tend to eat more.

The other major finding of this study was that food that had a higher nutrient density and lower energy density (fewer calories), such as whole foods, including vegetables, fruits, whole grain, beans, and nuts, led to a lower intake of overall calories, regardless of the time of day they were eaten.

If you eat empty calories from refined foods (such as doughnuts and sweet rolls) and sugars, you will tend to eat more overall. In other words, while breakfast alone is good, a good breakfast (one that is slowly absorbed, such as a meal containing protein—eggs, nut butters, a protein shake, or whole grains with nuts) is even better and perhaps more effective in preventing metabolic fluctuations later in the day. Maybe that's why studies have shown that besides being emotionally resilient, eating breakfast is one of the few things that is consistently correlated with longevity and a healthy weight.

Eating breakfast is just that, "breaking" the overnight "fast." Eating upon waking brings your blood sugar levels back to normal, kick-starts your metabolism, and sets you up to be on an even metabolic keel for the rest of the day. So break your fast every morning. It will make you healthier, give you more energy through the day, and help you lose weight.

Don't Eat Just Before Bed

As you remember from our discussion of the "sumo wrestler" diet, eating before bed is a guaranteed way to slow your metabolism and gain weight. The solution is simple: eat dinner earlier, and try not to go to bed for at least two to three hours after you eat. When you are asleep, all the hormone and messenger molecules that control your metabolism promote healing, repair, and growth. Also try to eat more lightly at dinner. Getting more of your energy needs in earlier in the day helps you lose and maintain weight loss. In fact, if you really want to prevent that slowdown at night, take a short stroll after dinner: it helps reduce your blood sugar and boost your metabolism.

Step 3: Enjoy Foods That Control Your Appetite and Avoid Foods That Send It Out of Control

The basic guidelines for eating in chapter 16 are designed into the Ultra-Metabolism Prescription. They will help you naturally and effortlessly control your appetite.

To help you create harmony in your hormones, turn on genes that make you lose weight, and control your appetite, you will want to concentrate on including whole, unprocessed foods that have good fats, a low GL, and a high PI. A list of specific foods is found on page 214.

To keep your appetite in balance, there are also some foods that you need to avoid wherever possible. These include some of the foods that you are going to eliminate in the preparation phase of the UltraMetabolism Prescription. They are worth repeating again here.

- Hydrogenated oils
- Refined vegetable oils
- Sugars
- Artificial sweeteners
- High-fructose corn syrup
- Flour products
- Refined grains
- Junk food
- Fast foods
- Processed foods

Step 4: Use Herbs to Optimize Your Hormonal Balance

Certain herbs have been shown to have an impact on appetite control. You may want to include the following herbal remedies in your program. Each of them has the potential to help control your appetite, and many of them can easily be used in the recipes that are outlined later in this book, or taken as supplements (see Appendix B).

* Ginseng
* Green tea
* Fenugreek
* Cinnamon

Step 5: Use Supplements to Control Your Appetite and Balance the Brain-Fat-Gut Cell Communication

The following supplements have been shown to help with appetite control. If you still have problems with this element of your weight loss program after instituting the earlier parts of this protocol, you might consider adding these supplements to your basic daily regimen (see chapter 16 and Appendix B for more information) if you have problems with appetite regulation or metabolic syndrome.

* Alpha-lipoic acid, a powerful antioxidant that improves glucose metabolism
* Gamma-linolenic acid (GLA), found in evening primrose oil
* PGX (PolyGlycopleX) or konjac root fiber, special fibers that absorb sugar and fat in the gut

Step 6: Consider Testing to Find the Causes of an Appetite That Is Out of Control

If you scored in the moderate to high range on page 86, you should seriously consider following up with additional testing that can help you confirm what your problems with appetite control might be and pinpoint what you need to do to change your condition. You should also seek professional medical help if you scored particularly high on the quiz.

I would start with the following basic tests. There are more advanced tests available as well, and information is available about them at www.ultrametabolism.com/tests.

Insulin and Glucose Tolerance Test

❖ A two-hour insulin and glucose response test is the best test for insulin resistance and metabolic syndrome, and I perform it at the least hint of belly fat or problems with sugar balance. Measuring insulin in addition to sugar is done by measuring your response to a sugar drink containing 75 grams of sugar (about 2 Cokes!). If your insulin level is high, your appetite will be out of control.

❖ Your fasting insulin level should be less than 8 mIU/ml, and a two-hour insulin test should be less than 30 mIU/ml.

❖ Your fasting blood sugar level should be less than 90 mg/dl and a two-hour sugar test should be less than 120 mg/dl.

Triglyceride and HDL Level Tests

❖ These are a measure of fats in your blood and the best indirect measure of insulin resistance. They are typically done as part of a cholesterol profile. But having a normal total cholesterol test can be misleading, because if the good cholesterol or HDL level is low (<40 mg/dl) it can make you appear to have a normal cholesterol reading, when in reality it is much worse than having a very high total cholesterol level with an equally high HDL. For example, if your HDL is 30 mg/dl and your total cholesterol is 180 mg/dl, your ratio of total to HDL cholesterol is 5, not good at all. If your total is 300 mg/dl and your HDL 100 mg/dl, your ratio is 3—much better (the ideal is less than 3). Your HDL should be over 60 mg/dl.

❖ Triglycerides (or the main type of blood fats) are increased with sugar consumption, particularly high-fructose corn syrup. A level over 100 mg/dl is considered high.

❖ *Note:* These measurements are accurate to assess metabolic syndrome only if you are *not* on a cholesterol-lowering medication.

The UltraMetabolism Prescription and Appetite Control

The UltraMetabolism Prescription is built on every one of the premises in this chapter. It will help you balance and control the brain–gut–fat cell interaction so they can work for you rather than against you. Making a few simple changes in your lifestyle and your meal composition, timing, and frequency, as well as reducing your glycemic load and increasing fiber by eating a whole-food diet, can shift your metabolism from hunger to satisfaction, from weight gain to weight loss, from feeling bad to feeling fabulous.

A Man of Uncontrollable Appetites

Appetite, a universal wolf.

—SHAKESPEARE

One evening after a lecture in New York, a man approached me to ask about seeing me as a patient. He was rotund, with a round, ruddy face, booming voice, and gentle manner. Everything about him was large—his appetites, his belly, and his heart.

He seemed tentatively curious about my work, and I was surprised when he showed up at my office. Samuel was nearly sixty years old, and his love of everything big was waning as he felt the encroachment of death. He described years of feeding his fat, eating a pint of ice cream every night before bed to keep his weight up.

At times he would have a change of heart and try to lose weight. He tried to lose weight as he had gained it. He made big attempts and tried some extreme diets, but the results were always the same: he gained back more than he lost.

Finally he realized that the big presence his 300-pound corpus gave him was not worth the infirmities he suffered. He was doggedly fatigued and short of breath at nearly every step, his nose was congested, and his legs were swollen. His skin was dry, and yeast was growing all over his body, fermenting on the sweetness of his skin.

He was not aware that he was diabetic and had dangerous cholesterol levels, angina, sleep apnea (a condition notable for snoring and dangerous stoppages of breathing at night), and a thyroid that was sluggish. He did notice that his leg hair was gone and that he had a more feminine appearance on his face and chest. Because of the estrogen being produced by his fat cells, his levels of female hormones were those of a woman.

He did not realize he was rusting inside, that his liver couldn't keep up with his body's detoxification demands, and that he was not effective at burning fat anymore. Once his large appetite for life had created his personality and he controlled it. Now it controlled him, and he felt powerless to stop it no matter how hard he tried to will the cravings away.

His testing showed severe sleep apnea, blockages in the blood flow to his heart, very small LDL and HDL cholesterol particles (the smaller the cholesterol particle, the more dangerous it is), dangerously high triglycerides, and very high insulin and blood sugar levels. His hormones were completely out of balance: his insulin was over 200 mIU/ml (normal <30) after a sugar drink, his estrogen level was high, his thy-

roid level was low, and his growth hormone level was very low. He also had food allergies and deficiencies of detoxification because of a fatty liver.

I told him that if he did everything I suggested, he would lose weight and feel better and all his symptoms would go away. Everything he had done, he had done to himself and could undo. I prescribed a simple program for him: the UltraMetabolism Prescription.

I had him eat a whole-food, unrefined diet without any restriction on calories or amount. I taught him the concept of nutrient density—foods with lots of nutrients and fewer calories, such as whole grains, beans, vegetables, fruits, seeds, and nuts. I had him start walking and slowly build up to more exercise. I gave him supplements to balance his blood sugar, antioxidants such as lipoic acid, and coenzyme Q10 (Co Q10) for his heart. I also gave him herbs to heal his fatty liver, fish oil, and extra fiber. He took a small dose of Armour thyroid to balance his thyroid, and we treated his sleep apnea with a special machine called CPAP that kept his airway open at night.

Enthusiastic though somewhat skeptical, he left my office determined to change. Three months later, I spoke to him again. He had lost 30 pounds, had more energy, and was beginning to look forward to exercise. His nasal congestion was gone and his fluid-filled, swollen legs were slimmer. All of his cravings were gone, he never felt hungry, and he found the program easy to follow.

Eight months later I saw him again and repeated his blood tests. I was shocked when he weighed in. He had lost 110 pounds without being on a strict deprivation diet. He simply changed his eating and his lifestyle. His diabetes was cured and his blood sugar level had dropped from 130 to 74 mg/dl (>126 mg/dl fasting denotes diabetes). His HDL and LDL cholesterol and triglyceride levels were normal without any medication. He was exercising vigorously three to four times a week and felt 20 years younger. He had, after a lifetime of uncontrollable appetites, finally found balance and health without suffering and continued taking pleasure in food.

Summary

❖ Your appetite control system is governed by a complex set of interactions between your gut, fat, and brain cells. You can gain some control over it using the six-step process outlined in this chapter.

CHAPTER 10

SUBDUE STRESS:

How Stress Makes You Fat and Relaxing Makes You Thin

Live Dangerously, Gain Weight

*E*ster *was an artist and loved life. She moved gracefully through it. Her children were her greatest passion—all beautiful and grown, smart and loving. Nearing sixty-five, Ester never struggled with her weight. Never much for formal exercise, she walked, gardened, and danced.*

Then one of her daughters moved to Israel during a time of particularly heated Israeli-Palestinian conflict. She began to watch CNN day and night, waiting, anxious, and expectant. Would her daughter be safe? Was the latest café bombing or suicide bomber in her neighborhood?

This went on for months and months despite attempts by her family to distract her. Her daughter would call and reassure her that she was happy and safe, but this did little to settle her anxiety. Her stress multiplied by the day, and so did her dress size. She gained 35 pounds watching CNN.

While the hours of television watched are directly correlated with weight gain, in Ester's case it was compounded by the state of alarm washing over her nervous system. It was as if she were living in a war zone herself and the hormones she produced while she was stressed out made her gain more and more weight. She didn't need to exercise more or change her diet; she needed to massage her nervous system, to quiet the molecules of alarm making her gain weight. She needed only to remove one thing from her diet: CNN. Once her daughter returned from Israel, her weight returned to normal.

Have We Become Caged Rats?

Imagine you are a rat immobilized by wires in a cage. It's bad enough being a lab rat, let alone being immobilized in a little wire box. You're completely stressed out. You're tied down in this cage, scared for your life, and as a consequence your body has triggered a special state of alarm—the primitive "flight-or-fight" response—where all of the molecules that scream "Danger!" to your body are flooding into your bloodstream.

As you lie in the cage, unable to move, a surge of molecules triggered by your brain starts streaming through your body, telling you that it is time to fight for your life or make a run for it. Your breathing increases, your blood pressure goes up, and oxygen shifts from the upper part of your body to your lower sectors preparing you to run. Your adrenal glands (small glands on top of the kidneys that control stress and fluid balance) release adrenaline and cortisol (the stress hormones) into your bloodstream, causing this chain of events to occur. This response also happens to increase blood fat, sugar, and insulin levels, preparing you for increased energy needs.

Studies of this nature have been done on lab rats, and you won't be surprised to hear that the chain of events described above is exactly what happens. The truly shocking finding, however, is what has been shown to happen to the weight of the poor rats that were tied down. Without any increase in calorie intake (actually, they ate less) or decrease in energy expenditure, they gained weight. (They actually expended more energy by struggling to get free.) They ate less, exercised more, and *gained* weight. That's right, they gained weight from stress alone![1]

Humans experience a similar "fight-or-flight" stress response. We have all felt that sudden rush in a moment of terror. The problem for humans is that we often feel stress, triggering this exact set of responses in our body, even when we aren't in any real danger. Because our psychology is vastly more complex than a rat's, we often feel stressed out about things that ultimately have very little impact on our physical safety.

How many of us have felt deep, unrelenting stress in the face of a situation that isn't an immediate danger to us? In fact, it is so commonplace in our culture that most of us fail to notice the effects of the chronic stresses we live with every day: demanding jobs, marital tension, lack of sleep, too much to do, and too little time to do it. I am sure the list goes on for many.

However, this mechanism, when triggered too often, leads to insulin resistance or metabolic syndrome, and as you remember from chapter 9, that's a condition that leads to weight gain. Chronically high levels of stress hormones circulating in the body have detrimental effects on humans similar to those seen in caged rats. This connection between stress and impaired metabolism is not well appreciated by doctors or the general public and is a recent breakthrough in scientific research.

The recognition that stress can lead to weight gain might be enough to make you jump off a cliff. How can you eat less, exercise more, and still gain weight? The truth is that it might actually make some evolutionary sense. Under any physical or psychological stress, the body is designed to protect itself, and one way it does that is to conserve weight. After all, if you're running from a predator, how can you be sure where your next meal is going to come from? Better to store calories for later use. This is just another example of the fact that the body we live with today is the product of a very old way of life.

Nonetheless, to fine-tune our metabolism and speak the language of our genes, we have to accept this truth. If stress can make you gain weight, you are going to have to learn to relax. But before we learn more about how and why to relax, let's have a look at how much stress you are coping with right now and whether it's preventing you from losing weight. The following assessment will help you.

HOW IS THE STRESS RESPONSE AFFECTING YOU?

Score 1 point each time you answer "yes" to the following questions by putting a check mark in the box on the right. See page 78 for a reminder on interpretation of your score.

	YES
Do you have low blood pressure?	☐
Do you get dizzy when you stand up?	☐
Have you been diagnosed with hypoglycemia?	☐
Do you have cravings for salt or sweets?	☐
Do you have dark circles under your eyes?	☐
Do you have trouble falling asleep and/or staying asleep?	☐
Do you feel groggy and not refreshed when you wake up?	☐
Do you experience mental fogginess or trouble concentrating?	☐
Do you get headaches?	☐
Do you get frequent infections (for example, colds)?	☐
Do you tire easily on doing any exercise or feel very fatigued after exercise?	☐
Do you often feel stressed?	☐
Are you simultaneously tired and wired?	☐
Do you have water retention?	☐
Do you have panic attacks or startle easily?	☐
Do you experience heart palpitations?	☐
Do you need to start the day with caffeine?	☐
Do you have poor tolerance for alcohol, caffeine, and other drugs?	☐

Do you often feel weak and/or shaky? ☐

Do you get sweaty palms and feet when you're nervous? ☐

Do you often experience fatigue? ☐

Do you often experience weak muscles? ☐

_____

If you don't want to write in this book, I've included this quiz in the companion guide that you can download by going to www.ultrametabolism.com/guide.

Identifying if you are stressed (and many people don't recognize the effects of stress on their health) is only the first step. Now it is important to understand what stress does to your body.

What Is Stress, and Where Does it Come From?

Stress is defined as a real or perceived threat to your body or your ego. It might be a rhinoceros chasing you (the story I recounted in chapter 1) or just a feeling of helplessness. It might be a psychological or social stressor such as depression, anxiety, grief, low socioeconomic status, divorce, loneliness, or unemployment. Or it might be a physical stressor: an infection, anything that triggers inflammation, exposure to cold temperatures, environmental toxins, pain, excessive exercise, smoking, alcohol, or stimulants.

Being overweight is a significant stressor. It has an impact on you both physically and psychologically. Being overweight leads to the production of more stress hormones, worsening the dangerous spiral of chronic elevations of those hormones, which cause depression, memory loss, bone loss, heart disease, cancer, and immune diseases. Besides just the physical stress of carrying around those extra pounds and the wear and tear on your joints, the hormonal and immune changes that occur with obesity lead to more rapid aging and degeneration.

In addition, you face many social and internal psychological pressures that come from being overweight: social derision, sexual problems, lack of ability to engage in enjoyable physical activities such as playing with your kids, shame about having to shop in stores for "big" men or women. The list goes on and on. Being overweight is an enormous stressor in itself.

To understand stress, we need to look at how stress is defined. Hans Selye, M.D., Ph.D., first coined the term "stress" in a paper in *Nature* in 1936 entitled "A Syndrome Produced by Diverse Nocuous Agents," in which he defined stress as "the nonspecific response of the body to any demand." Of course, Woody Allen and James Bond might have very different responses to the same stressor, but the key is the *perception* of stress. The danger doesn't have to be real; we just have to believe that it can affect us in real ways to set off all the physiological reactions that are associated with stress. Often the biggest stressors are not things or people, but *our thoughts* about them.

While most of you who pick up this book aren't actually in danger of losing your life on a regular basis, in modern society we are all regularly thrust into situations that make us feel stress. Consequently, we are constantly setting off a chain of reactions in our bodies that not only send us out of balance psychologically but actually make us gain weight. Let's look at how this happens.

The Brain-Gut Connection: How the Second Brain in Our Stomach Makes Us Fat

Your brain talks to your gut through a "second brain," or the enteric or "gut" nervous system. This is your *autonomic nervous system* (often referred to as the automatic nervous system), and it's made up of two parts. One part, the *sympathetic nervous system,* is responsible for the stress response. It slows things down, causes you to store fat, decreases your metabolism, increases your blood sugar, and leads to heartburn or reflux (where food moves up the digestive tract when it is supposed to go down) and constipation (where food stays where it is instead of moving through).

The other part is called the *parasympathetic nervous system* and is responsible for the relaxation response. It makes you digest your food, keeps things going in the right direction, increases fat burning, and lowers your blood sugar.

When you are feeling relaxed and safe, your body is programmed to digest and process food. When you are stressed and in danger, it is not the most opportune time to digest your lunch. Your survival needs focus all your energy on fighting or fleeing from stress or danger, not digesting food.

When you are stressed, the part that makes you fat (the sympathetic nervous system) gets turned on and all the signals go haywire: hunger increases, metabolism slows down, and you gain weight. Many people have made food their enemy; for them, eating is a stressful experience. They have to eat, but they worry about getting fat, so dining itself becomes a stressful experience for them. That means it isn't only the calories they consume but the stress they feel from consuming calories in the first place that's making them fat.

But that's only part of the story. When you are chronically stressed, you set off a series of hormonal imbalances that complicate the problem even more.

Stress: Hormones Out of Balance

When the brain is chronically stressed, your hormones get out of balance. Each of these imbalances can have a serious impact on your health and your ability to lose weight. Let's look at a few examples of how this can occur.

When you are introduced to a stressful situation, your body releases a hormone called *cortisol*. Cortisol is responsible for setting off all the physiological responses associated with stress. Studies have shown that when cortisol is released into the bloodstream you become less sensitive to leptin, the hormone that tells your brain you are full.

When this happens, you tend to eat more and crave more sugar. That means that your body doesn't only slow down your metabolism when you are stressed out, it actually tells you to consume additional calories. This makes perfect sense from an evolutionary perspective. When you are in danger, you want to eat all you can and store that energy so you're sure you will be able to face the dangerous times ahead. But in modern society, where we perceive danger when in fact we are perfectly safe, this mechanism is treacherous to our health. Here's a simple diagram that shows you how stress can lead to weight gain:

How Stress Causes You to Gain Weight

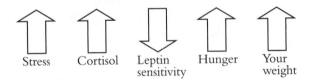

| Stress | Cortisol | Leptin sensitivity | Hunger | Your weight |

Innumerable studies have been done that show a direct link between stress and weight gain. Looking at a few of them should suffice to make you understand how vital this connection is to weight gain and how important it is to reduce stress to lose weight.

Research has shown that women with high levels of perceived and internalized racism had higher levels of cortisol and were more likely to be overweight and accumulate fat around their bellies.[2] Further, in a study of women, it was found that those with self-reported anxiety had higher levels of cortisol and cholesterol, lower levels of testosterone and thyroid hormone, and more weight around their middle than women who did not report such anxiety.[3]

Stress has also been shown to lower testosterone levels, which leads to muscle loss and fat accumulation. When men watched football and their team won, their testosterone level went up. When their team lost, a perceived stress, their testosterone level dropped.

Many other effects occur from being in a chronic alarm state that lead to a burned-out metabolism and eventually obesity. Growth hormone, testosterone, and HDL or good cholesterol levels drop, while insulin, blood sugar, and cholesterol levels and blood pressure all increase, all of which lead to weight gain.

Prolonged, unremitting stress[4] may lead to insulin resistance, diminished sex drive, and infertility. Additionally, you lose muscle mass, and your belly fat visceral adipose tissue (VAT) increases, making you into an "apple" shape. Your cholesterol, blood pressure, and triglycerides go up. You are more tired, but your sleep is restless. You get more and more depressed. The immune system gets turned up and the inflammatory response increases, as does the "rusting" process, or oxidative stress. Your thyroid gets sluggish, and your detoxification system gets overloaded. Without a way to counteract the effects of chronic stress, we are on a slippery downward slope to weight gain and ill health. This type of constant and unremitting stress counteracts all the 7 Keys to UltraMetabolism.

The VAT You Gain from Stress Makes You Even Fatter

But the problems that stress causes your waistline are worse yet. Not only does your body slow down your metabolism and tell you to eat more, but the fat that you gain from being stressed out actually communicates negative messages to your body.

The body has a stereotypical way of reacting to stress, wherever it comes from and whatever form it presents itself in. As you read above, when you feel stress, your brain triggers an alarm state that sets off a series of chemical responses that slows down your metabolism and starts to store calories. A good portion of this is the visceral adipose tissue (VAT).

The old idea that fat is just a storage depot for energy you need during starvation is fast falling away. Fat cells are now considered an endocrine organ, a part of your hormonal communication system. Not only are fat cells an active endocrine organ, sending messages out to the rest of the body to regulate weight, metabolism, stress hormones, and inflammation; they are even wired directly to your autonomic nervous system. That means that your brain is telling them what to do without your slightest awareness of or control over the process.

One of the main hormonal messengers that your fat creates is cortisol. Remember that cortisol is your major stress hormone. It's responsible for setting off the entire physiological chain of events associated with stress. This includes slowing down your metabolism.

As you may be able to see by now, this whole process very quickly turns into a vicious cycle, where stress and weight gain feed each other constantly. When you get stressed out, your body releases cortisol, which inhibits your responsiveness to leptin and slows down your metabolism. As a consequence, you gain VAT. Then this VAT starts sending more cortisol into your bloodstream, starting the whole process over again.

The bottom line? Being stressed makes you fatter, the extra fat you put on around the middle (VAT) produces more stress hormones, leading to more fat storage, leading to more stress hormones, and more fat—and on and on the vicious spiral turns. The only way to stop it is to *relax*.

The Role Your Genes Play: The Gene-Stress-Fat Link

What's more, some individuals have a genetic variation that make it difficult for them to process cortisol properly. Researchers have found that for some people under chronic stress, the feedback loop to the brain that normally shuts down cortisol production is impaired. These people have variations in their genes (*polymorphisms*) that result in an inherited inability to put the brakes on the stress response. As a consequence, they are perpetually caught in a cycle of stress weight gain and more cortisol production. These people tend to gain even more weight under stress than the average person does.

But before some of you suggest that obesity is mostly genetic, consider this study. Twenty pairs of identical twins who differed by more than 37 pounds in weight were studied. The overweight twins had higher levels of the stress hormones adrenaline and cortisol and poorer-quality sleep, drank more alcohol, and had higher perceived levels of stress. There was no difference in their genes, only their stress levels.[5]

The good news is that even people who have a genetic predisposition to gain weight can have an immense impact on what their genes tell their bodies to do by giving their genes the right kind of information. If you are genetically predisposed to have problems coping with excess cortisol in your system, I recommend the same thing to you that I do for anyone facing a chronic stress problem: *relax*.

The Problems with Stress Don't Stop There: Loss of Rhythm, Loss of Sleep, Metabolic Burnout, and Night-Eating Syndrome

In situations where stress is persistent, you also lose the normal circadian, or daily, rhythm of your hormones. Consequently, certain hormones go up when they should be down and down when they should go up. Ultimately, your system just burns out, a state referred to as *metabolic burnout*.

The hormone cortisol normally rises in the morning to wake you up, increase your appetite, and give you energy for the day. At night it normally falls and growth hormone and melatonin levels rise, helping you sleep and repair your body. With metabolic burnout, this normal rhythm disappears, causing even more weight gain.

In some cases, people start suffering from "night-eating syndrome," a condition that leads to a decrease in appetite in the morning, increased hunger and eating at night, and difficulty losing weight. These people typically have high levels of cortisol. In one study of such individuals, however, a mere one week of relaxation training resulted in lower levels of cortisol, hunger, and food intake at night.[6]

Sleep deprivation is another major source of stress in our modern world. Americans, on average, sleep two hours less now than they did forty years ago. What effect does this have on metabolism and weight? A group of researchers found that depriving healthy men of sleep led to increases in *grehlin,* the hunger hormone, and decreases in *leptin,* the satiety hormone. This increased their hunger and craving for calorie-dense, high-carbohydrate foods. After suffering from many nights of sleep deprivation while working in the emergency room, I can tell you that this is true!

Relaxing Makes You Thin

All of these disturbances in your metabolism lead to weight gain. Furthermore, everything in your body is interconnected, and factors that lead to increased insulin production, such as high-glycemic foods or a big meal, also cause stress and higher cortisol levels,[7] perpetuating another vicious cycle.

> Chronic stress negatively affects all 7 Keys to UltraMetabolism.

Chronic stress negatively affects all the 7 Keys to UltraMetabolism. It is also one of the major factors in the development of metabolic syndrome. You can see that reducing stress plays a major role in losing weight and getting healthy. But how do you do that? The answer is simple: learn to relax.

As mentioned earlier, your stress response is part of your autonomic nervous system. The part of this system that sets off the stress alarm is your sympathetic nervous system. When you are stressed, the brain sends signals through this alarm system (your sympathetic nerves) to your fat cells. This slows down your metabolism, reduces fat burning, makes you more insulin-resistant, and hence leads to weight gain. This is a part of your body that you have no conscious control over no matter how hard you try, so there is no way to stop this process once it's begun.

The good news is that you *do* have control over the part of your nervous system that helps you relax: the parasympathetic system. When you relax, you turn off the genes that cause you to gain weight and turn on the genes that cause you to lose weight. When you relax, your metabolism is turned up, your fat burning increases, you become more insulin-sensitive, and consequently you lose weight. In other words, relaxing is a great weight loss strategy!

The problem is that relaxing isn't a natural state. Most of us are used to being stressed out, but we have no idea what it means to relax. Thus you need to train yourself to relax. There is a way to do it, and the rest of this chapter will teach you how.

You can train your body and mind to relax by following this six-step protocol:

- **Step 1:** Identify and reduce the causes of stress.
- **Step 2:** Practice active relaxation.
- **Step 3:** Eat stress-reducing foods and avoid stress-producing foods.
- **Step 4:** Use herbs to reduce your stress.
- **Step 5:** Use supplements to reduce your stress.
- **Step 6:** Consider measuring your stress response with testing.

If you are having problems with this UltraMetabolism key and you are dealing with too much stress, you can turn the tables and start coping with your stress by following these steps. Doing so will customize the Ultra-Metabolism Prescription for your own particular needs, allowing you to turn on genes that cause you to lose weight and turn off the ones that have been making you gain weight. If you scored high on the quiz in this chapter, try to seek out the testing mentioned in Step 6 as well as professional medical help. Let's look at exactly how you can make this six-step process work for you.

Step 1: Identify and Reduce the Causes of Stress

The first step in reducing your stress level is identifying what stresses you out (stressors) and eliminating them. In some cases this might be easier said than done. If your boss is a major stressor for you, eliminating him probably isn't the answer. Nonetheless, there are many stressors that you can get rid of if you chose to.

Generally speaking, there are two kinds of stressors: psychosocial and physical. Psychosocial stressors are social or psychological events that cause you stress. Physical stressors are physical problems that cause your body to stress out. You may not be consciously aware of all the things that are stressing you out in these two areas right now, but it is important for you to take an inventory of the ones you are aware of so you can begin to do something about them.

To start with, I would suggest you take out a piece of paper and make a list of the things in your life that are causing you stress. You might fold this piece of paper in half and on one side write "psychosocial stressors" and on the other "physical stressors." Once you have done this, simply list the things in each column that stress you out.

Here are a few common psychosocial stressors to consider: job, relationship, financial situation, kids, psychological disorders (depression, anxiety, etc.), low self-esteem, the state of the world (the international political situation, problems in your neighborhood, etc.).

Here are a few physical stressors to consider: being overweight or obese, chronic illness, allergens, toxins, sugar and high-fructose corn syrup, saturated and trans fats, chronic infections, alcohol, tobacco, drugs.

Once you have made a list of your various stressors, try to think of ways that you might eliminate them. In some cases, this is easy. For example, if you feel as though you drink a little too much in the evenings and it might be a physical stressor for you, drink less. In other cases, it will be more difficult. If you are dealing with a chronic psychological or health issue, for example, it will probably take some energy to get that worked out.

But always remember, you *can* take action to reduce the amount of stress in your life.

Step 2: Practice Active Relaxation

Even if you can't eliminate all the stressors in your life, there are ways that you can encourage your body and brain to relax intentionally when you are faced with stressful situations. What follows is a simple breathing exercise you can use to consciously relax and recommendations for sauna use. But before you get started, there are a few recommendations I would like to make about how and when to practice relaxation.

First, you should practice relaxation consistently if you can. It is important to do this both for your weight and for your health.

Try to schedule two to three thirty-minute sessions of relaxation exercises per week if you can. Your weight depends on it. Besides, once you get started you will find it feels better than you might expect.

Make sure that you are secluded and won't be interrupted during your relaxation periods.

Before you get started on any of the exercises, you might want to take a moment to breathe and center yourself. Just take a few deep breaths or use the deep-breathing exercise on page 122 to start this process.

In addition, you shouldn't hop right up and get back to work after your relaxation time is over. This will basically undo most of the progress you've made. Take some time to come out of the exercise and back to your surroundings. Move slowly after you are finished. Feel your body. Look around the room. Feel how relaxed you are. Then go back to your life.

Now that you have some background on how to relax, let's get to the exercises themselves. We will start with the deep-breathing exercise because it is foundational and easy to do.

Activating the Relaxation Response

Engaging in activities that turn on what Herbert Benson, M.D., of the Harvard Mind-Body Institute calls the *relaxation response* can be a great antidote to chronic psychosocial or physical stress. There is a variety of techniques that achieve the same state of relaxation, including meditation, yoga, autogenic training (a technique for relaxation based on awareness of body sensations), progressive muscle relaxation (systematically tensing and relaxing different muscle groups), guided imagery (purposefully focusing and guiding thoughts and mental images), and hypnosis. Laughing, listening to beautiful music, and making love all turn on the same response. A recent study in *Alternative Therapies* found that those who practiced yoga regularly lost more weight than those who didn't.[8] This research demonstrates the significant role that stress management can play in weight loss.

You can choose a technique that conforms to your own beliefs. The relaxation response can be experienced by practicing any of the following in the proper manner:

- ⁜ Meditation
- ⁜ Certain types of prayer
- ⁜ Autogenic training
- ⁜ Progressive muscular relaxation
- ⁜ Guided imagery
- ⁜ Hypnosis
- ⁜ Lamaze breathing exercises
- ⁜ Yoga
- ⁜ Tai qi quan
- ⁜ Qi gong

I have provided a few simple ways on the following pages so you can start relaxing right away—you can use your breath or even a sauna!

Deep Breathing

Start by getting into a comfortable position. If you're at home, sit in your favorite chair or lie down somewhere comfortable. If you're at the office, sit comfortably on your chair with your feet on the floor, your back straight, and your hands on your lap.

You may want to unbutton the top button of your pants or loosen your shirt a little. Do whatever you need to make yourself comfortable. When you have done this, close your eyes and begin the relaxation process.

Slowly bring your attention to your breathing. Notice how you are breathing. Are you breathing in through your nose or in through your mouth? Are you taking shallow breaths or full, deep ones? Where can you feel your breath in your body? Is it in your chest or in your belly?

Once you have a sense of your breathing, slowly begin inhaling through your nose, then exhaling slowly through your nose. Don't rush yourself. Don't hyperventilate. Just calmly breathe in and out through your nose.

As you do this, try to see if you can breathe deeply into your abdomen. Place your hand on your belly, and as you breathe make your hand rise and fall. You can place your other hand on your chest if you wish. As you start breathing deeply into your abdomen, you should feel your belly extend more than your chest.

Feel your body begin to relax as you breathe deeply. Now when you exhale, think the word *relax*. Inhale. Then exhale. *Relax*. Inhale. Exhale. *Relax*.

You can continue this process for five to ten minutes if you wish. Simply focus on breathing into your abdomen and telling yourself to relax. As you do so, feel what happens to your body. Let any places where you feel tension unwind. Let your body feel like a loose rope. Relax physically.

When you are ready, slowly bring yourself back to the present moment. Look around at the room you are in. Gradually go about your day with this new relaxed attitude.

A Short Version of Deep Breathing

You can use this technique in a short version anytime you feel tense. You may have some experience with this already. When you feel stressed out, you may take a deep breath and let out a big sigh to release some of your tension. Consciously turning this into an opportunity to relax can be a powerful experience.

When you are stressed out, simply close your eyes, and take five to ten deep breaths into your abdomen, and silently say the word *relax* as you exhale. Be aware of the relaxed feeling that comes over your body when you do this. Allowing yourself to unwind just for that moment can be a powerful way to keep stress at bay.

Take Saunas or Steam Baths

Taking saunas or steam baths has also been proven very effective in reducing the stress response and creating balance in the autonomic nervous system. They improve circulation, help with weight loss, balance blood sugar, and improve detoxification.

In addition, saunas have been shown to reduce complications and improve cardiac performance in heart patients. When the autonomic nervous system is in chronic stress mode, the heartbeat becomes less variable. Normally there is a subtle variability between beats. If there is more variability in the beat-to-beat rhythm, your heart and nervous system are healthier. The least healthy heart rhythm has the least variability—a flat line. Saunas or steam baths increase the variability and health of your nervous system.[9]

While the exact mechanism is not clear, it is likely due to its effects on calming the nervous system, relaxing the muscles, and increasing circulation. It may also be due to direct effects on the brain's control center, the hypothalamus, which is affected by temperature. It also helps eliminate toxins that activate the stress response.

To take a sauna or steam bath, follow these guidelines:

❖ Find a local sauna.
❖ Start slowly with five to ten minutes at a session.
❖ Build up to thirty to sixty minutes with cooldown periods or a cold shower in between.

✢ Stay well hydrated during the sessions; take a glass bottle of water into the sauna or steam bath.

✢ Consider purchasing an infrared sauna (www.sunlightsaunas.com) for home use.

✢ Be sure to take a good multimineral supplement when doing sauna treatments (if you are not doing so already) to replace the minerals lost through sweating.

Though saunas have been used to treat heart disease and diabetes, if you have any chronic health condition, be sure to check with your physician before starting sauna therapy.

Step 3: Eat Stress-Reducing Foods and Avoid Stress-Producing Foods

The basic principles of healthy eating outlined in chapter 16 are designed to reduce stress and heal your metabolism. Follow those guidelines for foods to enjoy and foods to avoid.

There are a number of foods that will help you relax and reduce the impact of chronic stress and others you need to avoid to keep from causing any more damage to your body. Nutritional support is very important in subduing stress. Improving blood sugar and insulin control by reducing your intake of refined sugars and carbohydrates (high-glycemic-load carbohydrates) and increasing your intake of omega-3 fats, fiber, and specific vitamins, such as the B-complex vitamins, zinc, vitamin C, and other antioxidants, is very important. Elevated blood sugar and insulin levels feed into the vicious cycle of adrenal gland stimulation and the production of the stress hormones.

Nutrients are also important in helping the sensors for the steroid hormones, such as cortisol, to function properly. These include B6, B12, and folic acid. These vitamins are also important in detoxifying excess adrenaline and cortisol so they can be eliminated from the body.

Others that may also be helpful are vitamin B5 or pantothenic acid and vitamin C. Potassium and magnesium are also balancing for the adrenal glands. They are found in dark green leafy vegetables and many fruits, vegetables, and whole grains. Increasing your intake of omega-3 fats from wild fish and whole ground flaxseeds or flaxseed oil can reduce inflammation and its negative effects on the hypothalamic-pituitary-adrenal axis.

Antioxidants, also found in fruits and vegetables, provide additional defense against the effects of stress in the body. Stress causes oxidation or "rusting" (see chapter 12), so antioxidants such as vitamins E and C as well as coenzyme Q10 and lipoic acid may be helpful. A good multivitamin and mineral supplement can take care of most of these needs (see chapter 16).

Step 4: Use Herbs to Reduce Your Stress

Certain herbs help you adapt to stress better. If you want to add yet another dimension to your stress reduction program, try these herbal remedies. You can reduce the overactivity and increase the resilience of your system by using adaptogenic (named because they may help you adapt to stress) or balancing herbs in combination or singly. They include:

- Ginseng
- Rhodiola
- Siberian ginseng
- Ashwagandha
- Licorice

Step 5: Use Supplements to Reduce Your Stress

If you are interested in adding supplements to your stress reduction regimen, there are a few that have been shown to help undo the effects of physical stressors. You might try adding the following to your basic supplement regimen:

- B-complex vitamins
- Magnesium
- Vitamin C
- Zinc

Step 6: Consider Measuring Your Stress Response with Testing

Specific tests can help identify the health of your stress response and whether it is over- or underactive and can be helpful in suggesting the right treatment.

Adrenal Stress Index

This involves four separate saliva tests for cortisol at four different times of the day. A number of labs perform this test, which helps you identify if your stress response is functioning normally, you are on overdrive, or you are burned out. Each finding may require a different treatment.

24-Hour Urinary Cortisol Test

This is a measure of stress hormone in the urine. When this is very elevated, it is considered a disease (Cushing's syndrome), but if it is mildly or moderately elevated, it is a great indicator of an overactive stress response. Testing this and finding high levels has been one of the best ways to show people why they can't lose weight and why they need to relax.

IGF-1 Test

This is a measure of growth hormone, which goes down whenever stress or cortisol goes up. Growth hormone is important for building muscle and keeping you young.

Stress and the UltraMetabolism Program

Unlocking the keys to UltraMetabolism requires seeing the whole body as a system. Stress affects everything, and almost everything we do can cause stress, from sleep deprivation to eating sugar to feeling overwhelmed by life. Working diligently to address all of these components can make an enormous difference in your health and your weight over time.

Sleeping It Off

Joseph was in his early 50s and weighed 306 pounds. He was a corporate executive who was making great strides in his career. He had worked incessantly to climb the corporate ladder. But as he climbed, his weight climbed.

He was an opportunistic feeder: whenever there was an opportunity, he ate, with little planning or forethought. Suddenly he was hungry, and he ate whatever was available—a bag of pretzels, a soda, or some other low-fat, high-refined-carbohydrate snack. The vending machine at work was his ally. Coffee was taken as a slow infusion to keep him going throughout the day.

When I first saw him, he was on a low-fat, low-salt vegetarian diet to try to control his weight, high blood pressure, and high cholesterol level. Doctors had plied him with a cocktail of medications to try to help: Lipitor for his cholesterol and three blood pressure medications. Still, despite treatment, he had metabolic syndrome and was impotent.

As he grew larger around the middle, he also grew more tired. Unable to sit and work at his desk without falling asleep, he had a standing desk fashioned for his computer and worked standing for most of the day. He developed varicose veins.

His wife was in the room during our first interview. Does he snore? I asked. Does he snore? she replied. His snoring was so bad he had been exiled to sleep in another bedroom. It seemed that his weight, prediabetes, high blood pressure, and cholesterol all stemmed from not sleeping. Chronic sleep deprivation disturbs all the hormones that control weight and increases hunger and cravings for sugar and starchy carbohydrates, leading to all the complications of high blood pressure, cholesterol, and diabetes. After he was fitted with a special device to treat his sleep apnea, stop his snoring, and finally give him a restful night's sleep, he lost 50 pounds. I also asked him to stop his low-fat, low-salt, junk-food vegetarian diet and put him on a diet of whole, real foods. He was no longer impotent, his triglycerides dropped from 259 to 80 mg/dl, and his HDL level went from 38 to 59 mg/dl (both in the ideal range). His blood sugar and insulin levels returned to normal, and he was finally able to work sitting down without falling asleep.

SLEEP APNEA

Sleep apnea is a disorder in which a person stops breathing during the night, perhaps hundreds of times, usually for periods of ten seconds or longer. In most cases the person is unaware of it, although sometimes he or she awakens and gasps for breath. Apnea is usually accompanied by snoring. People who have sleep apnea may not even be aware of the condition, but inevitably it causes daytime sleepiness, weight gain, and heart and lung disease, as well as high blood pressure and insulin resistance. It often goes undiagnosed. If you are a snorer (just ask your partner) and overweight, you should have a sleep study performed by your doctor.

Summary

❖ Stress releases cortisol, a hormone that sets off into your bloodstream a chain of chemical events that makes you gain weight. You can reverse this process by practicing relaxation.

CHAPTER 11

COOL THE FIRE OF INFLAMMATION:

Hidden Fires That Make You Fat

A Body on Fire: The Fat-Inflammation Connection

*J*ill was one of my "whole-listic" patients. She was 52 when she first came to see me with a whole list of complaints. She seemed to have one of every disease: chronic fatigue, fibromyalgia, irritable bowel syndrome, inflammatory bowel disease, rheumatoid arthritis, reflux, chronic sinusitis, hypothyroidism, high cholesterol level, insulin resistance, food allergies, asthma, and a rare immune deficiency disease. And she was overweight, which really troubled her.

She had always been thin and had gained more than 43 pounds during the course of her illness. She was on a full pill bottle of medication every day—one pill for every ill—prescribed by innumerable doctors, all trying to help cure a disease. For years she and I struggled together trying everything I knew, doing every test, looking for every clue and cause, and we found many: mercury, food allergies, hormonal imbalances, nutritional deficiencies, and stress, all of which contributed to her generalized inflammatory state. She had more inflammatory diseases than anyone I had ever seen. How could one person have so many inflammatory problems?

Her gut, her joints, her stomach, her sinuses, and her lungs were all inflamed. We treated all the problems, and she would improve slightly, only to relapse again. Then one day she came to see me at the end of her rope, after months of antibiotics for chronic sinusitis, increasing weight gain, intolerable fatigue, and joint pains. She was even getting monthly intravenous injections of antibodies from her doctor to help her immune system, because her own was not working properly. She was willing to try anything.

I advised her to go on a strict anti-inflammatory detoxification program for two weeks—only low-allergy rice protein (typically sold in concentrated powder form), rice, and steamed vegetables and fish. She had a remarkable and dramatic recovery from nearly all her symptoms. It was then I realized that she might not have been completely eliminating what we had found years before to be a problem: gluten.

She had tried to eliminate it, but it always slowly slipped back, resulting in relapse after relapse. I sent her a very strong e-mail telling her to completely, absolutely, 100 percent, no exceptions eliminate gluten from her life for three months. She got the message. Gluten intolerance (from eating wheat, barley, rye, and oats) and food sensitivities or allergies cause inflammation, fluid retention, and weight gain. Gluten is one of the most common undiagnosed causes of inflammation that promote weight gain.

The next time I saw her was when I was giving lectures at a theater in New York. I heard someone call my name: "Mark, is that you?" I turned to see a woman in a beautiful evening gown looking radiant and svelte. It took me a minute or so before I realized it was Jill, 35 pounds lighter and completely healthy.

All her inflammatory and autoimmune diseases had been caused by celiac disease. Celiac disease is characterized by intolerance to a grain protein called gluten, found in wheat, barley, rye, oats, spelt, and kamut. It is an allergic condition that affects many people, though few have been diagnosed with it because it takes so many different forms, one of which is weight gain or the inability to lose weight.

She had finally gotten off gluten completely, and her world had changed. When I saw her that night, her husband turned to me and said, "Thank you for giving me my wife back."

What Is Inflammation?

What is inflammation, and what does it have to do with being overweight?[1] Remarkable new research links obesity and inflammation. Being overweight promotes inflammation and inflammation promotes obesity in a terrible, vicious cycle. More than half of Americans are inflamed, and most of them don't know it. Getting to the root of inflammation and cooling it off is key to reducing the obesity epidemic and your own waist size.

What is inflammation, anyway? Most of us are familiar with inflammation. The classic signs are pain, swelling, redness, and heat—as with a bad sore throat or infected hangnail. This is a good thing, as it fights foreign invaders of all types.

Inflammation is part of the body's natural defense system against infection, irritation, toxins, and other foreign molecules. A specific cascade of events occurs in which the body's white blood cells and specific chemicals (cytokines) mobilize to protect you from foreign invaders.

But sometimes the natural balance of the immune system, which produces just enough inflammation to keep infections, allergens, toxins, and other stresses under control, is disrupted. The immune system shifts into a chronic state of alarm or inflammation, spreading a smoldering fire throughout the body. This fire in the heart causes heart disease, in the brain causes dementia and Alzheimer's disease, in the whole body causes cancer, in the eyes causes blindness, and, as we are just discovering, in our fat cells causes obesity.

While on the one hand this inflammatory process is protective, it can go awry, not only in individuals with inflammatory diseases such as arthritis but in otherwise healthy individuals whose lifestyles and/or environments expose them to substances the body perceives as irritants, such as low-grade infections from gum disease, food allergens, toxins, and even inflammatory foods such as sugar and animal fat.

Likewise, while inflammation is sometimes obvious, such as when an injured area becomes swollen, red, and warm to the touch, what science is teaching us is that inflammation can occur much more quietly and insidiously. It can occur silently without any symptoms. It is even emerging as a major cause of heart disease, diabetes, cancer, Alzheimer's disease, and aging in general. It is also connected to weight gain. Inflammation is a silent killer, and unless it is adequately dealt with it can have disastrous effects on your weight and your health.

Anything that causes inflammation can make you gain weight, and any weight you gain can cause more inflammation. The most common cause of systemic inflammation is our modern diet (sugar, animal fat, and processed food, or the high-glycemic-load diet most Americans are eating) and lack of exercise. Other things contribute but to a lesser extent, such as food (particularly gluten) and environmental allergens, infections, stress, and toxins.

To help you get a sense of whether or not you are suffering from chronic inflammation, take the following assessment exam.

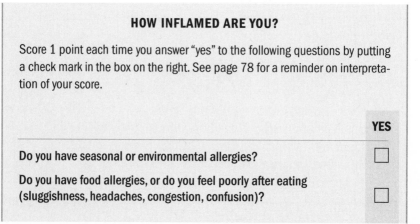

HOW INFLAMED ARE YOU?

Score 1 point each time you answer "yes" to the following questions by putting a check mark in the box on the right. See page 78 for a reminder on interpretation of your score.

	YES
Do you have seasonal or environmental allergies?	☐
Do you have food allergies, or do you feel poorly after eating (sluggishness, headaches, congestion, confusion)?	☐

Do you work in an environment with poor lighting, chemicals, and poor ventilation? ☐

Are you exposed to pesticides, toxic chemicals, loud noise, heavy metals, and toxic bosses and coworkers? ☐

Do you get frequent colds and infections? ☐

Do you have a history of chronic infections such as hepatitis, skin infections, canker sores, and cold sores? ☐

Do you have sinusitis and allergies? ☐

Do you have bronchitis or asthma? ☐

Do you have dermatitis (eczema, acne, rashes)? ☐

Do you suffer from arthritis (osteoarthritis/degenerative wear and tear)? ☐

Do you have an autoimmune disease (rheumatoid arthritis, lupus)? ☐

Do you have colitis or inflammatory bowel disease? ☐

Do you have irritable bowel syndrome (spastic colon)? ☐

Do you have problems such as ADHD, autism, mood, or behavior problems (actually part of a family of problems called neuritis)? ☐

Do you have heart disease, or have you had a heart attack? ☐

Do you have diabetes, or are you overweight (BMI greater than 25)? ☐

Do you have Parkinson's disease or have a family history of Parkinson's or Alzheimer's disease? ☐

Do you have a significant amount of stress in your life? ☐

Do you drink more than three glasses of alcohol a week? ☐

Do you exercise less than thirty minutes three times a week? ☐

If you don't want to write in this book, I've included this quiz in the companion guide that you can download by going to www.ultrametabolism.com/guide.

If you are inflamed for any reason, it is very important to find the cause and reduce the inflammation, not just for the purpose of weight loss but because it is a major cause of all the major degenerative diseases of modern civilization: heart disease, dementia, diabetes, and cancer. Now that you know how inflamed you are, you can take action to turn off the fire that's burning in your body. This chapter will teach you how to do that.

The Body's Web:
Determining the Causes of Inflammation

The fact that medicine is divided into specialties has nothing to do with how the body is really organized. It was a convenience to create specialties such as endocrinology, cardiology, immunology, and so on when we knew very little about how the body works. In fact, it works like a web in which everything is connected to everything else.

While using the 7 Keys can help to understand what is going on, they too are only a way of describing a very complex interconnected system that is our body. While we tend to focus on single systems or particular symptoms as reflections of problems with certain parts of the body, the truth is that your body is an intricate web.★ Each part of it affects every other part. Inflammation is an excellent example of this.

If any one of the other Six Keys to UltraMetabolism is out of balance, it can cause you to become inflamed. That means that everything you have learned up to this point in the book and everything you have yet to learn are contributors to inflammation. An unbalanced diet, a poor eating lifestyle, stress, lack of exercise, oxidation, an out-of-balance thyroid, and a toxic, fatty liver all increase the likelihood that you will be inflamed.

Diets high in saturated or animal fats and trans fats also increase overall levels of inflammation,[2] as can eating too many fake foods and too many calories in general. Eating a diet that has a high glycemic load has been shown to increase inflammation powerfully. In fact, the latest advances in obesity research have identified inflammation as a main culprit leading to weight gain, especially when related to imbalances in blood sugar caused by insulin resistance, or prediabetes (and, as you know from chapter 10, insulin resistance is often set off by a high-glycemic-load diet). In fact, in one study, being inflamed increased the risk of diabetes by nearly 1,700 percent.[3,4]

Get a C-Reactive Protein Blood Test
to Check for Inflammation

If you are concerned about your level of inflammation, I would strongly recommend that you talk to your doctor about testing your levels of *C-reactive protein* (CRP). CRP is a protein found in the blood, and it's the

---

★ For physicians or those interested in the science behind this new medical paradigm, I recommend *The Textbook of Functional Medicine*. See www. functionalmedicine.org for more information.

best indicator we have of a heightened state of inflammation in the body. You may also be able to order the test for yourself from one of the resources listed at www.ultrametabolism.com/tests. The more information you have about how inflamed you are, the better position you will be in to respond to this condition.

In the meantime, let's look a little more closely at how inflammation occurs in the body and explore some ways you can start counteracting it right now.

The Fire in Your Belly: Fat Cells Stoking the Flames

Your own fat cells, the ones around your middle, are often the biggest source of inflammation. This is because the fat cells (called *adipocytes*) do more than just hold up your pants or provide stored energy for some future date when you may be starving. They produce hormones, such as *leptin,* which reduces appetite; *resistin,* which makes you more insulin-resistant; and *adiponectin,* which makes you more insulin-sensitive and lowers your blood sugar. In addition, they produce the hormones estrogen, testosterone, and cortisol. Recently we have discovered that fat cells also produce inflammatory molecules (remember the cytokines) such as IL-6 and TNF-alpha. As you can see, your fat cells are busy controlling appetite, hormonal balances, and inflammation.

Adipocytes, the medical term for fat cells, produce *adipocytokines,* another fancy term for inflammatory molecules or cytokines that come from fat cells. You might get a little confused here. Do fat cells produce hormones or cytokines? Are they part of the endocrine system or part of the immune system? The answer is all of the above!

The molecules produced by your fat cells wreak havoc on your metabolism by increasing inflammation, increasing your appetite, slowing fat burning, and increasing stress hormones. They exist in abundance when your system is out of balance from too much stress, too much sugar, too many trans fats, or exposure to an overload of toxins, allergens, or infections.

These molecules possess strange names, such as IL-1-beta, IL-6, TNF-alpha, resistin, leptin, adiponectin, and more. That fat around your middle, your VAT, makes a fire in your belly that spreads throughout your system by sending out these inflammatory messenger molecules throughout your body. And fire anywhere else in the body creates more fire in the belly, creating a vicious cycle of inflammation, oxidative stress, and metabolic changes that leads to weight gain and the metabolic syndrome, or prediabetes.

The bottom line: Fat cells promote inflammation, which leads to more fat cells, which promotes more inflammation until you are very inflamed and very overweight. Getting rid of inflammation helps you lose fat, and losing fat helps you get rid of inflammation.

PPAR, The Switch for Inflammation and Weight Gain—The Link Between Your Diet and Your Genes

Central to understanding the inflammation-metabolism-weight connection is a group of receptors that are abundant on the surface of the nucleus of your fat and liver cells called the *PPAR family* (alpha, beta, and gamma).[5] PPARs are tiny docking stations on your cells that communicate with your DNA to turn your metabolism up or down. They also happen to control inflammation. Certain foods (particularly certain kinds of fats) turn on these PPAR receptors, and certain other foods turn them off.

They explain how inflammation causes weight gain and show us how we can use natural anti-inflammatories to lose weight.

The food you eat talks to your genes. Specifically, it speaks to the PPAR family of receptors, telling them to make you lose or gain weight, cool off inflammation, or increase the fire. Of course, we need both the ability to create inflammation to fight off infections or injury and the ability to reduce inflammation.

That balance is the key to your health. One of the inflammatory molecules produced by our fat cells when we eat too much sugar, too much saturated fat, or just too many calories is called *tumor necrosis factor alpha* (or TNF-alpha). This binds to and blocks the PPARs inside your cells. This inflammatory molecule slows your metabolism, makes your body resistant to the effects of insulin, and causes weight gain. Eating poor-quality foods (sugar, trans fats, saturated fat, etc.) that don't fit our evolutionary needs triggers your body to release inflammatory messages that prevent critical parts of your metabolic system from working, thus crippling your metabolism and causing you to gain weight. This isn't just a theory; we now know how these foods speak to your genes.

Some of the most interesting research of recent years has revolved around how the same foods that aid in weight loss (omega-3 fats found in fish and flaxseed oil and vegetables and fruits rich in antioxidants and phytonutrients) also inhibit inflammation.

Turning Off Inflammation

Many studies show that changes in diet, lifestyle, and exercise can reduce inflammation. A diet higher in fiber, olive oil, and low-glycemic-load

continues on page 136

DRUGS OR FOOD? WORKING AGAINST OR WITH THE BODY

Enormous amounts of research dollars are being spent to find new drugs that will turn PPAR receptors on and off, thus creating a "magic" weight loss pill. A new class of drugs called TZDs (thiazolidinediones), which are used for diabetes to improve insulin sensitivity and reduce inflammation, have promise in this area.

As usual, the focus is on a synthetic compound that will interfere with the body's function in some way. Medical researchers believe drugs are our only weapons against disease; therefore, they are the only tools that are researched, even when food or lifestyle or natural products can be more effective.

But drugs are not the solution. PPAR receptors were not put in our cells so that one day a drug company could make billions of dollars by finding a molecule that turns them on. They were put there by design to regulate our energy metabolism using signals we were exposed to every day: food and exercise! Fish oil binds to PPAR receptors to do the same thing these drugs are trying to do, and it does it better than the medications and at a substantially reduced cost (a TZD drug costs $164 a month, a bottle of fish oil pills $10 to $20 a month).

In fact, a recent study in the *British Medical Journal* countered another study that promoted a "polypill" that could be given to all people to prevent heart disease. Giving everyone a combination pill containing a low dose of aspirin, a statin, a beta-blocker, an ACE inhibitor, and folic acid, the first researcher maintained, could significantly reduce heart disease.

The new study showed that a "polymeal" was even better; 4 ounces of wild fish (e.g., salmon), 5 ounces of red wine, 3.5 ounces of dark chocolate, 2.5 ounces of almonds, 400 grams of fruits and vegetables (about 1 pound, or the equivalent of 2 2/3 apples), and 1 clove of garlic a day could reduce all heart disease by 75 percent and increase life expectancy by seven years by sending the right messages to your genes. And there are no side effects.

Drugs Have Side Effects; Food Does Not

Have you ever taken the time to read the package insert that accompanies any prescription medication? It rather makes one wonder what's worse, the existing condition or the possible side effects from ingesting "the cure." There will always be physiological consequences

(continued)

from taking a drug. There are inescapable biological laws of cause and effect. Medical and pharmacology schools instill in their students early on that, whether obvious side effects appear or not, all medications are toxic to the body to varying degrees and often result in unintended consequences.

Medications have a role in medicine, but they should be the last resort in treating disease, not the first. Drugs block or interfere with our biology in some way. These are the antidrugs, the inhibitors, and the blockers in our medicine cabinets: anti-inflammatories, antidepressants, antibiotics, beta-blockers, calcium channel blockers, ACE inhibitors, and so on. They are designed to work not with the body but against it and therefore have significant potential side effects and risks.

The first question should be not "What is the best drug for this problem?" but "How was the body designed, and how can I work with it to help return it to a state of balance and optimal function?" The body is brilliantly designed to function well when all the right ingredients are provided. Your genes receive messages from the environment every moment. Food and the molecules contained in food contain all the information to switch on genes that restore normal functioning—*without any side effects.* Special messages in food act on the same places in your cells as medication, but instead of blocking something, they facilitate normal physiology and turn messages that restore balance and health and optimal metabolism on or off.

carbohydrates, lower in saturated fat and cholesterol, and high in omega-3 fats reduces inflammation and improves insulin sensitivity.[6] Even just increasing fiber intake lowers inflammation and C-reactive protein levels.[7]

Dr. David Jenkins, from the University of Toronto, has shown that eating a combined diet of soy foods, soluble or viscous fiber (glucomannan or konjac), almonds and plant sterols (plant fats found in small quantities in fruits, vegetables, and nuts) lowers both cholesterol and inflammation as much as the popular statin drugs now being heavily promoted to lower both cholesterol and inflammation.[8] Soy foods have also been found to lower inflammation.

In addition, eliminating the causes of inflammation and adding herbal remedies and supplements to your UltraMetabolism Prescription can dramatically improve the inflamed state of your body. To turn off the fire in your body, use this six-step protocol:

✢ **Step 1:** Eliminate the factors that cause inflammation.
✢ **Step 2:** Direct your genes to turn off inflammation signals.
✢ **Step 3:** Eat foods that reduce inflammation, and avoid foods that cause it.
✢ **Step 4:** Use herbs to reduce your inflammation.
✢ **Step 5:** Use supplements to reduce your inflammation.
✢ **Step 6:** Test for inflammation and its causes.

If you are having problems with this UltraMetabolism key and your body is on fire, you can turn the tables and start controlling your inflammation by following these steps. Doing so will customize the UltraMetabolism Prescription for your own particular needs, allowing you to turn on genes that cause you to lose weight and turn off the ones that have been making you gain weight. If you scored high on the quiz in this chapter, try to seek out the testing outlined in Step 6 as well as professional medical help. Let's look at exactly how you can make this six-step process work for you.

Step 1: Eliminate the Factors That Cause Inflammation

We have to find the factors that increase inflammation and get rid of them. You can take all the anti-inflammatory drugs you want (or eat all the fish oil or chocolate nibs you want), but if you don't get rid of the cause you will simply be covering up the symptoms.

It's like the old adage from my mentor, Sidney Baker, M.D.: "If you are standing on a tack, it takes a lot of aspirin to make you feel better. The treatment for standing on a tack is removing the tack." The treatment for inflammation comes on your plate and in your shoes. What you eat and how much you exercise are the most important factors governing inflammation.

Finding the causes of inflammation is not always easy. The most common and obvious causes are our diet and being sedentary. But there are many factors, and at times specialized testing is needed to find hidden causes.

Dietary factors such as excess sugar, refined carbohydrates, saturated and trans fats, or just too many calories can also cause inflammation. Sometimes the cause may be a hidden infection, something you eat or breathe that you are allergic to, or an environmental toxin.

Stress will also make you inflamed. And though just sitting around doing nothing also causes inflammation, regular exercise is one of the best-known anti-inflammatories on the planet.[9] Multivitamins are also a great natural inflammation-fighting tool.[10]

By identifying the sources of inflammation—sugars and refined high-glycemic-load, rapidly absorbed carbohydrates; saturated and trans fats; lack of exercise; gluten; food allergies; mold in damp basements or moldy bathrooms, or hidden in walls; a hidden infection such as a virus, parasite, or bacterium that doesn't cause immediately obvious symptoms; or a medication you are taking—and getting rid of them you can stop chronic inflammation.

Sometimes this requires detective work, testing, and working with an experienced doctor, but the results for your weight and your health will be worth the effort.

Step 2: Direct Your Genes to Turn Off Inflammation Signals

To turn off the inflammation in your body, you need to send the right messages to your genes. Lack of exercise (see chapter 13) and stress (as we learned in chapter 10), in addition to a processed, high-glycemic-load diet (see chapter 9), all tell your genes to produce more inflammation.

The key message of this book is that *food communicates with your genes.* Food affects your weight (and everything else) by sending messages to burn or store fat. It does this through something called *transcription factors.* Transcription factors are regulatory proteins that initiate the transcription of certain genes upon binding with DNA. That means these little proteins actually make certain parts of your genetic code readable while ignoring other parts of it. Depending on which parts of your DNA these transcription factors decode, your body gets messages to put on weight or burn fat more rapidly. These transcription factors are regulated in part by what we eat. This is how food "talks to" your genes and tells them to either turn on genes that cause you to become inflamed and overweight or turn on genes that cause inflammation and weight to be reduced.

One of the most important transcription factors is called *nuclear factor kappa B* (NF-KB). This little molecule creates much damage. If activated by emotional stress, toxins, or free radicals or certain toxic, inflammatory, or allergic foods, it can unleash the production of more than 125 inflammatory molecules—a fireball of inflammation that affects your whole system.

The good news is that antioxidants from food and supplements can turn this off. We will come back to our new friend, NF-KB, in the following chapter on oxidative stress, but I bring it to your attention now because oxidative stress is another major cause of inflammation, and it is oxidative stress that activates NF-KB, starting the terrible cycle of inflammation.

Eat the Right Fats

One of the best ways to send the right messages to your genes, reduce in-flammation, and burn fat is to eat the *right* fats: the omega-3 fats (EPA and DHA) found in fish oil.[11] A new fat has been discovered that also controls our energy metabolism and inflammation through the PPAR receptors: OEA (oleoylethanolamide)[12] (commonly found in cocoa butter, dark chocolate and cocoa nibs; see below for more information on these).

Turning on PPAR receptors with fish oil, antioxidants, or chocolate shuts down NF-ΚB and thus reduces inflammation and oxidative stress. In addition to fish oil and cocoa butter, which is found in dark chocolate or cocoa nibs, antioxidants also turn on the PPAR receptors. Therefore antioxidants also help you become more insulin-sensitive, burn more fat, and reduce inflammation. Antioxidants in food and in supplements may be helpful in regulating your weight, as you will learn more about in chapter 12.

Eat Dark Chocolate

Further, foods that are high in phytonutrients have a wide range of antiox-idant and anti-inflammatory actions

Some of these phytonutrients, called *flavonols,* are found in fruits and vegetables such as berries, grapes (hence the benefits of red wine), tea, and cocoa. Chocolate[13] (the dark kind) contains phytonutrients called *polyphe-nols.** These are natural antioxidants and anti-inflammatory molecules that cool off inflammation and can help protect you against obesity. That's right, I said it. Chocolate can be a weight loss food!

Before you go out and stock up on Snickers, there are some caveats. First the chocolate must be free of added saturated fats and rich in cocoa. Cocoa is the plant that chocolate is made from, and it is where the antioxidant and anti-inflammatory polyphenols come from.

This combination of properties can be found only in special kinds of dark chocolate. So when you go out and stock up on chocolate, take the following factors into consideration: The chocolate should have only mini-mal amounts of added sugar, and it should have as much cocoa in it as pos-sible. Many of the dark chocolates that are on the market today specify the

---

* Substances found in many plants that give some flowers, fruits, and vegetables their color. Polyphenols have powerful antioxidant and anti-inflammatory ac-tivity. They are abundant in berries, green tea, and cocoa.

content of cocoa on the label. You should aim for at least 70 percent cocoa content. In addition, keep in mind that even this special kind of dark chocolate should be eaten in moderation, about two to three ounces a day.[14] Don't start eating fifteen chocolate bars a day and think it will improve your health!

You can also try cocoa nibs—roasted, unprocessed, whole cocoa beans. They are crunchy, delicious, and full of polyphenols and OEA, a special fat that helps you burn fat (see Resources for sources of cocoa nibs).

Eat Fruits and Vegetables

There are many other sources for phytonutrients than chocolate and cocoa. You can get many anti-inflammatory phytonutrients by eating plant foods that have a high phytonutrient index (remember, this is the measure of the overall amount of healing plant chemicals in plant foods). If you don't eat enough plant food, you will never get enough of these healing nutrients. Eating whole, unprocessed real foods that have a low GL and high PI is another important way to reduce inflammation in your body. If you don't eat high-quality food, you are going to become inflamed and fat.

For instance, studies have shown that the same amount of calories from fast-food meals creates more inflammation and oxidative stress than those from meals containing fruits and vegetables, which contain those important phytonutrients. So once again, it is not just the calories you eat but where they come from that is important in losing or maintaining weight, creating a healthy metabolism, and reducing inflammation. A diet high in phytonutrients is critical if you want to lose weight. The UltraMetabolism Prescription is just such a diet.

Move Your Body

Exercise is something our ancestors never did. When I traveled to rural China in 1984, no one was jogging. They were too tired from carrying buckets of water and sewage, from working in the fields all day and sawing boards from logs by hand. Sitting next to Paul Ridker, M.D., the pioneer from Harvard who has proven the link between inflammation and heart disease, at a conference on nutrition, genetics, and inflammation, I asked him why we were all inflamed in the twentieth century. He replied that we are all doing a lot less in terms of moving around and using our bodies (currently known as exercise) than our ancestors were adapted to doing. In fact, reducing inflammation is probably the main way that exercise helps to prevent heart disease. Before you panic, just remember that *any* moving around counts. In chapter 13 we will learn about exercise and the mitochondria and how to make exercise as painless as possible.

Subdue Stress

You are now an expert in stress from reading the previous chapter. You know where it comes from and how to manage it. What you may not realize is how much stress can increase inflammation, nor how much relaxing and reducing stress can reduce inflammation. Study after study has shown that even in diseases such as asthma and rheumatoid arthritis, not to mention obesity, stress reduction has real, powerful anti-inflammatory benefits.

Step 3: Eat Foods That Reduce Inflammation, and Avoid Foods That Cause It

Your diet is the most important factor in helping you reduce inflammation. The UltraMetabolism guidelines for an optimal diet (see chapter 16) are an anti-inflammatory diet by design. In addition to typical foods that increase inflammation, such as sugar, processed foods, and trans fats, certain food allergies, sensitivities, and gluten sensitivity can be common sources of problems with weight and metabolism.

Foods to Avoid That Cause Inflammation

In addition to following the basic guidelines outlined in chapter 16, specific foods may be a particular problem for some people.

- *Food allergens* (the most common are wheat, dairy, eggs, corn, soy, and peanuts). Elimination and reintroduction of these foods, as in Phase I of the UltraMetabolism Prescription, is the best way to find out if you react to these foods, though testing can be a useful guide as well.
- *Gluten.* I pull this out separately because it affects 1 percent or more of the population (millions of people) and is an often undiagnosed cause of inflammation. Testing can confirm if this is a problem. Gluten containing grains includes wheat, rye, barley, spelt, kamut, and oats.

Step 4: Use Herbs to Reduce Your Inflammation

Using anti-inflammatory herbs can have remarkable results. Try adding these herbs to your regimen for inflammation reduction either in your diet or as supplements.

- Capsaicin (from cayenne pepper)
- Green tea
- Ginger
- Quercitin (in fruit and vegetable rinds)
- Turmeric (the yellow spice found in curry)
- Cocoa

Step 5: Use Supplements to Reduce Your Inflammation

Taking the following supplements is an additional way to decrease the fires of inflammation. Try adding them to your standard supplement regimen of multivitamins and minerals and fish oil, which by themselves are powerful anti-inflammatories.

- Probiotics
- Enzymes (bromelain and other proteolytic enzymes)

Step 6: Test for Inflammation and Its Causes

If you scored in the moderate to high range on the quiz at the beginning of the chapter, you can pinpoint the causes of your inflammation and confirm that it is indeed a problem for you by taking the tests described on the next page. More advanced testing is also available. In addition to the tests, I strongly encourage you to seek professional medical assistance if you scored high on the quiz.

High-Sensitivity C-Reactive Protein (hs-CRP) Test

This is the best test for inflammation. It measures the general level of inflammation but does not tell you where it comes from. The most common reason for an elevated C-reactive protein is metabolic syndrome or insulin resistance. The second most common is some sort of reaction to food—a sensitivity, a true allergy, or an autoimmune reaction, as occurs with gluten. The special tests to follow can help identify food problems and gluten reactions. One caution about C-reactive protein: If it is high (and high is anything over 1.0), you can be sure you are inflamed, but if it is normal, you can't be 100 percent sure there isn't some smoldering inflammation somewhere (if only medicine were an exact science!).

IgG Food Sensitivity Tests (Antibody Tests Against Food)

Though these tests are still controversial, well-controlled studies[15] have shown them to be helpful in identifying problem foods and that removal helps inflammatory problems. I have found them to be imperfect though helpful guides in locating trouble foods.

ELISA/RAST IgE Tests (Tests for Mold and Environmental Allergies)

This is the classic allergy blood test done by allergists for acute allergies. However, chronic mold exposure and allergies, particularly from "sick" buildings, are a growing problem that leads to inflammation and ill health.

Gluten Intolerance/Celiac Disease Tests

These tests help identify various forms of allergy or sensitivity to gluten or wheat. The diagnosis can sometimes be difficult. You may want to refer your doctor to an important review paper, "Narrative Review: Celiac Disease: Understanding a Complex Autoimmune Disorder" by Drs. Armin Alaedini and Peter Green from Cornell and Columbia University (*Annals of Internal Medicine,* February 2005, pages 289–298). It sheds light on how best to diagnose this common condition that shows up in many ways but is often overlooked. Below are the most common blood tests used to identify this problem, which can cause a host of inflammatory diseases, from autoimmune diseases to obesity to dementia and even cancer.

- IgA antigliadin antibodies
- IgG antigliadin antibodies
- IgA antiendomysial antibodies
- Tissue transglutaminase antibody (IgA and, in questionable cases, IgG)
- Stool antigliadin IgA and tTG (www.enterolab.com), a home test kit for wheat or gluten allergy
- HLA DQ2 and DQ8 genotyping, gene testing for the celiac or gluten sensitivity gene

Inflammation: The Hidden Cause of Weight Loss and Most Chronic Diseases

The importance of finding the source of and treating inflammation cannot be overstated. It is something most physicians were not trained to do. As you have learned, there *is* a way to identify if you have inflammation with a C-reactive protein test. Then you can find the causes in your diet and environment and correct them. You can also do many things to help reduce inflammation by including anti-inflammatory foods, herbs, and supplements in your diet, exercising, and relaxing. Your weight will come down as you cool the fire of inflammation, and the rest of your health will improve at the same time.

The Fifty-Five-Pound Menopausal Weight Gain

Shauna arrived at my office in despair. After going through menopause, her life began coming apart at the seams. Crying in my office, she recounted how she had gained sixty pounds in the last six years, and her marriage was falling apart. She felt depressed, frustrated, tired, and unable to lose weight. She wanted to feel better and wanted to exercise and eat better, but she felt exhausted all the time, and whenever she did manage to exercise, she would fatigue very quickly.

It had all started six years earlier. Shauna went to her doctor for some help with hot flashes. He prescribed Premarin, the most common form of estrogen (derived from pregnant mare's urine, hence the name), which was the standard treatment at the time. Shortly thereafter she began to retain fluid, gain weight, and develop high blood pressure.

To treat the high blood pressure, her doctor started her on another drug called Tenormin, a beta-blocker. This type of medication slows both heart rate and blood pressure and can lead to depression, fatigue, decreased libido, and more. One of its worst side effects is that it makes it harder for cells to metabolize sugar. As a result, her body produced more insulin to get the sugar into the cells. This, in turn, had a domino effect, causing more weight gain, higher blood pressure, more sugar cravings, and more fatigue. So after having been healthy most of her life, Shauna put on 55 pounds over two to three years, becoming depressed and prediabetic—all from a doctor's visit for hot flashes.

When we did some tests on Shauna we found something very interesting. I knew she had two grandmothers with diabetes, and while I expected that she was prediabetic or insulin-resistant, we were shocked to find that her system was on fire. We measured her C-reactive protein and found something astounding. Normally it should be less than 1.0 mg/L (milligrams per liter); over 3.0 mg/L is considered very high. Her level was 22.0 mg/L, which indicated that an incredibly high amount of inflammation was running rampant throughout her body. The hidden inflammation interfered even further with her metabolism and led to more insulin resistance and weight gain. But what was the cause of the inflammation?

It was the hormone she was taking, Premarin. That hormone increases inflammation, as well as fluid retention and blood pressure, and causes weight gain, leading to the need for high blood pressure medication.

Two years after Shauna began the UltraMetabolism Prescription, she had lost all of the 55 pounds she had gained. When asked what she thought made the most difference, she replied, "Getting off the Premarin and beta-blocker. As long as I took those medications, I couldn't lose weight." After she discontinued the drugs that increased inflammation and interfered with her metabolism, she was finally able to stop the vicious cycle of fatigue, sugar cravings, and weight gain and was able to eat better and exercise more.

The hot flashes stopped, and her blood pressure became normal. Her C-reactive protein dropped from 22.0 to 1.8 mg/L, and her good cholesterol level went up 30 points. Now, two years later, she is happy, her marriage has found its way back to more stable ground, and she looks ten years younger.

Note: Be sure never to stop any medication without your doctor's supervision.

Summary

- ❖ Inflammation is connected to every major health threat we face today. It is in large part responsible for heart disease, cancer, diabetes, Alzheimer's disease, arthritis, allergic diseases, and all the autoimmune disorders.
- ❖ Inflammation and obesity are intimately tied together. They create a vicious cycle where inflammation leads to weight gain and weight gain leads to additional inflammation.
- ❖ You can reverse this process by adjusting your diet and your lifestyle.

CHAPTER 12

PREVENT OXIDATIVE STRESS OR "RUST":

Keep the Free Radicals from Taking Over

Finding the Source of Rusting

*I*first met Florence on one of her many trips to Canyon Ranch, where I was co–medical director for nine years. She was always an eager patient, soaking up information about eating well and exercise. She was perhaps 5 to 10 pounds overweight, and nearing her mid-40s. Though she had premature menopause and bone loss, she was otherwise healthy.

The stress of a tough divorce and being a single working mother made her mildly depressed. She couldn't tolerate the Prozac her doctor gave her, so we worked with alternatives. Shortly after her divorce she moved into a new house. It was then that things went haywire. She developed chronic fatigue and low thyroid function and gained more and more weight until she was at least 60 pounds heavier than when I had first met her.

Despite her exercising and trying to eat well, her liver became fatty, her blood sugar, insulin, and cholesterol levels went way up, and her stress hormones soared. Her appetite increased, and she had trouble controlling her sugar cravings. She went on disability from her job as a teacher. Something was seriously wrong. Her tests showed many things, including a high level of oxidative stress or "rusting," as shown by blood and urine tests for rancid fats and damaged DNA.

This rusting process wreaks havoc on the metabolism, primarily through cell damage, and we couldn't seem to get it under control. Things spiraled. We treated her for food allergies, got rid of mercury in her system, tried antioxidants and many supplements. She would improve a little but wasn't responding as I thought she should. Then her daughter developed juvenile rheumatoid arthritis that got better whenever they went on a vacation. The light went on—it was something in her environment.

We checked for molds and hit the jackpot. An environmental engineer found multiple toxic molds in her environment. We found antibodies to those exact same molds in her blood, as well as antibodies to the mold toxins.★ Her body was repeatedly try-

---

★ *Epicoccum nigrum, Penicillium notatum, Polaria pullans, Rhizopus nigricans,* and *Stachybotris,* as well as antibodies to *Alternaria, Geotrichum,* and *Candida.*

ing to fight off these toxins, and over time these efforts increased oxidative stress, in-
flamed her liver and her body as a whole, and led to weight gain because they dam-
aged the weight control centers in her brain.

Her leptin levels were high—a sign of leptin resistance, which prevents the normal
appetite control mechanisms from working. And she had very low levels of one of the
major brakes on the appetite, a neurotransmitter called alpha-melanocortin–
stimulating hormone, or α-MSH.

Her insurance company agreed to have the house torn down and build another for
her. She moved, most of her symptoms improved, and she finally lost forty-three
pounds over four months.

Rusting Leads to Weight Gain

Free radicals are taking over; that's why we are gaining weight. But free radicals are not a left-wing political group left over from the 1960s. They are
both a central cause and a result of obesity.

Free radicals, or reactive oxygen species (ROSs), cause a condition called
oxidative stress, or *oxidation*. Oxidation is commonly seen in the visible
world as the rusting of a car or a sliced apple turning brown. It is also responsible for the wrinkles on your face when you have been exposed to too
much sunlight over the years. The problem is that rust not only exists in the
visible or outside world, we actually rust on the inside. And this contributes
to impaired metabolism, weight gain, and aging. Think of this as wrinkles
on the inside.

Let's see how rusted you are. Take the following self-assessment to get a
sense of how out of balance this key to UltraMetabolism is for you.

ARE YOU RUSTING?

Score 1 point each time you answer "yes" to the following questions by placing
a check mark in the box on the right. See page 78 for a reminder on interpretation of your score.

	YES
Are you fatigued on a regular basis?	☐
Are you sensitive to perfume, smoke, or other chemicals or fumes?	☐
Do you regularly experience deep muscle or joint pain?	☐
Are you exposed to a significant level of environmental pollutants or chemicals at home or at work?	☐

(continued)

Do you use tobacco products? ☐

Are you exposed to secondhand smoke? ☐

Do you drink more than three alcoholic beverages a week? ☐

Are you exposed to sunlight or ultraviolet light
(tanning booths) more than one hour a week? ☐

Do you exercise less than ½ hour three times a week? ☐

Do you take prescription, over-the-counter, and/or recreational drugs? ☐

Would you describe your daily stress level as high? ☐

Do you eat fried foods, margarine, or high-fat foods? ☐

Do you eat less than five to nine servings (½ cup each)
of deeply colored vegetables and fruits a day? ☐

Do you tend to overeat often? ☐

If you don't want to write in this book, I've included this quiz in the companion guide that you can download by going to www.ultrametabolism.com/guide.

Oxidation, like every other problem presented in this book, is something you can take control over so that it doesn't stand in the way of your weight loss program. This chapter will teach you how.

Reducing Rust = Reducing Weight

Oxidation is a basic chemical reaction found everywhere in nature. It is a natural part of your biology but one that becomes a significant problem when it runs out of control. When something is oxidized, it is damaged by oxygen. But not the form of oxygen you breathe, O_2, with which we are familiar. Oxidation is caused by a lonely form of oxygen called just O.

Oxygen molecules like to be paired. They like to have two electrons. Free radicals are a form of oxygen that contains only one oxygen molecule, so they run around your body looking for an electron to steal from another molecule—sort of like stealing someone's spouse. Then the molecule that it borrowed an electron from is damaged (or oxidized). And it in turn runs around looking to find an electron to steal. This process leaves a wake of destruction in its path: damaged DNA, damaged cell membranes, rancid cholesterol, stiff arteries that end up looking like rusty pipes, and wrinkles.

Oxidized tissues and cells don't function normally. Part of the malfunction is the promotion of weight gain and a damaged metabolism.

Following is a list of symptoms common to people who are dealing with a high level of oxidation. If you are suffering from any of the following, you may want to consider reducing the rust in your body.

- Fatigue
- Poor mental function and cognition
- Lowered resistance to infection (perhaps frequent colds or sinus infections)
- Muscle weakness
- Muscle and joint pains
- Digestive problems (reflux, irritable bowel syndrome, ulcers)
- Anxiety
- Depression
- Headaches
- Hypoglycemia (low blood sugar; symptoms may include dizziness, anxiety, sweating, or nausea and may occur if more than an hour or two elapses between meals or snacks)
- Allergies
- Irritability
- Dizziness

Oxidation is balanced by antioxidants in a process called *reduction*. The role of antioxidants—both those obtained from diet or supplements, and those produced naturally in the body—is to reduce the number of free radicals, thus curbing damage to your cells and metabolism.

If you reduce oxidized molecules in your body, you will make yourself healthier and reduce your weight, not to mention reduce your wrinkles. The dietary balance between antioxidants (found in colorful plant foods and supplements) and oxidants (produced mostly by nutrient-poor, refined,

RUST AND INFLAMMATION: ANOTHER VICIOUS CYCLE

Oxidative stress and inflammation are intimately connected. Rusting leads to inflammation and inflammation leads to rusting. Anything that reduces oxidation or rusting reduces inflammation, and anything that reduces inflammation reduces oxidation. They are both the cause and effect of each other in our bodies. This process of rusting and inflammation can run out of control, leading to aging, heart disease, cancer, and dementia, as well as impaired metabolism and obesity.

and processed foods with too many calories, such as soft drinks and baked goods) ultimately controls the whole process of aging and chronic disease, including weight and metabolism. But where do these free radicals come from and how can we control oxidative stress?

Too Many Calories and Not Enough Antioxidants: Why the Free Radicals Are Taking Over

In humans, most of the oxidation or production of free radicals occurs during the natural process of metabolism, turning calories or energy from food and oxygen from the air into energy usable by the body. This is normal, and normally limited because we historically ate food full of antioxidants, such as berries and nuts.

This is no longer the case in modern society, where we subsist on large quantities of food with empty calories. (Ever wonder why scientists and health professionals consider them "empty"? Because they contain no antioxidants or nutrients.) This balance between oxidants and antioxidants is central to health because it controls many cellular messages, including the ones related to your weight.

The way antioxidants and oxidants control our weight is through their effect on our genes. They signal the genes to enhance or impair our metabolism through their effect on receptors such as PPAR and transcription factors such as NF-KB (our friends from the previous chapter). Antioxidants protect you against weight loss, inflammation, and diabetes; oxidants or free radicals trigger a cascade that leads gene messages to be turned on that promote weight gain, slow your metabolism, increase inflammation, and cause diabetes.

The single most important controllable factor regulating the oxidative stress in the body is our diet. Consuming too many calories and not enough antioxidants from colorful plant foods (including the polyphenols in dark chocolate) results in the production of too many free radicals.

Free radicals are generated in the energy factories of our cells known as *mitochondria*. Mitochondria are built to convert calories and oxygen into energy the body can use: adenosine triphosphate (ATP). Free radicals are a by-product of this conversion process.

Our cells contain a total of 100,000 trillion mitochondria, which consume 90 percent of our oxygen intake. This oxygen is necessary to burn the calories we eat in food. But the free radicals are produced as a by-product of combustion, much like exhaust that comes out of the tailpipe of your car. We have our own antioxidant systems to protect us, but these systems are easily overwhelmed by a toxic, low-nutrient, high-calorie diet. The antioxidants we make (called superoxide dismutase, catalase, and glutathione peroxidase) are dependent on essential dietary nutrients to help them work well, including zinc, copper, manganese, vitamin C, and selenium.

We are in trouble with free radicals for two reasons. First, we eat a diet that causes an escalation in free radicals because it has too many empty calories and not enough antioxidants. Second, our reduced nutrient intake in the form of vitamins and minerals limits our ability to make our own antioxidants, which need nutrients such as zinc and selenium to function properly. That's why these self-manufactured antioxidants are not enough to protect us.

But there is a way to increase the amount of antioxidants in your body and reduce the effects of rusting. You can customize the UltraMetabolism Prescription by following this five-step protocol:

- **Step 1:** Eliminate the causes of oxidation.
- **Step 2:** Eat foods that reduce rusting and avoid foods that cause it.
- **Step 3:** Use herbs to reduce your oxidation.
- **Step 4:** Use supplements to reduce your oxidation.
- **Step 5:** Consider testing for oxidation in your body.

If you are having problems with this UltraMetabolism key and your body is rusting, you can turn the tables and start controlling the oxidation process by following these steps. Doing so will customize the UltraMetabolism Prescription for your own particular needs, allowing you to turn on genes that cause you to lose weight and turn off the ones that have been making you gain weight. If you scored high on the quiz in this chapter, try to seek out the testing outlined in Step 5 as well as professional medical help. Let's look at exactly how you can make this five-step process work for you.

Step 1: Eliminate the Causes of Oxidation

You must eliminate the factors in your diet, lifestyle, and environment that cause oxidation. What follows is a list of common ways that you can eliminate hidden factors that are causing you to rust.

- **Avoid overeating.** Excess caloric intake contributes to oxidative stress. When you eat too many calories, you create more "exhaust" in your metabolism. These are the normal waste products produced by the mitochondria when burning foods. The waste products *are* the free radicals. Our body's ability to cope with those excess free radicals is overwhelmed, and they run out of control, damaging your metabolism.

- **Avoid charbroiled foods.** These foods contain polycyclic aromatic hydrocarbons (PAHs). This is the black crispy part of grilled food, which generates more free radicals in the body.
- **Avoid excess sugar and refined carbohydrates.** They are "empty calories." They actually burn hotter and generate more free radicals.
- **Avoid excess alcohol.** It can lead to an increase in cytokines (inflammatory molecules) and free radicals. So while 5 ounces of wine may be beneficial, 15 ounces may be harmful. In the case of alcohol, if a little is good, more is not better.
- **Reduce your exposure to toxins and petrochemicals** (in plastic storage containers, tap water, and pesticides) **and heavy metals** (such as in contaminated fish). Eliminating these elements from your environment and your diet is important. Drink filtered water and eat organic foods, fruits, vegetables, and animal products raised without hormones, antibiotics, or pesticides.
- **Minimize your exposure to ionizing radiation.** This includes ultraviolet radiation from excess sun exposure, X-rays, and radon—check your house for this.
- **Reduce your exposure to tobacco smoke (first- or secondhand).** Cigarette smoke has more than four thousand toxic chemicals that lead to accelerated oxidative stress.
- **Reduce air pollutants.** Place HEPA or ULPA filters (available commercially) in your home. Allergens and particulate matter from industrial pollution increase oxidation and irritate your immune system.
- **Avoid excessive or insufficient exercise.** If you exercise vigorously more than sixty minutes a day on most days, that is considered excessive. On the other hand, less than thirty minutes five times a week isn't enough. There is more information on how to exercise in chapter 13.
- **Get at least seven to nine hours of sleep each night.** Sleep deprivation is another stress, and any stress increases oxidative stress, promoting weight gain.
- **Treat chronic infections.** Hidden or chronic infections promote oxidative stress.
- **Reduce internal sources of oxidative stress.** Dietary imbalances from processed foods and high-glycemic foods, too many calories, and abnormal levels of gut bacteria or yeast are all forms of oxidative stress.
- **Improve your liver and gut detoxification.** As you will learn

(see chapter 15), if the liver is overloaded with toxins from food (too much sugar or trans fats) or toxins from the environment such as pesticides or mercury, it becomes inflamed and produces more free radicals. Loving your liver is key to keeping the free radicals under control.

- **Reduce your exposure to fungal toxins.** These include environmental and internal molds and fungi. Many people work in sick buildings or have moldy basements or bathrooms. The molds produce toxins that increase oxidative stress and free radicals in our bodies. It is important to identify environmental sources of molds.
- **Reduce stress.** Cortisol and stress both increase inflammation. Any kind of stress to the body—physical, such as extremes of heat and cold, injury or trauma, over- or underexercise, and psychological stress all generate the same response in the body: inflammation and oxidation.
- **Improve your breathing and oxygenation.** Provide more oxygen to your tissues and cells through deep breathing and yoga. Even though oxygen seems to be the problem, we all know we can't live without it (except for about four minutes before we die). By bringing more oxygen to the tissues through deep breathing and yoga, you can flush toxins and damaging free radicals and inflammatory molecules from your body.

Step 2: Eat Foods That Reduce Rusting and Avoid Foods That Cause It

Dietary antioxidants (vitamins, minerals, polyphenols, and phytonutrients) are the major factors that protect your cells from excess oxidative damage and impaired metabolism. A nutrient-poor, calorie-rich, high-glycemic-load, antioxidant-deficient diet increases oxidative stress through cell messengers such as NF-KB[1] and triggers signals that make you gain weight.

This is why antioxidants are a key part of the body's system to lose weight and keep it off. If you are rusting, you are also getting fatter.

To reduce oxidative stress, you must begin by increasing your intake of a variety of antioxidant-containing, colorful plant foods,[2] rich in phytonutrients such as flavonoids and polyphenols (found in teas, red wine, and cocoa).[3,4]

Your plate should look like an impressionist painting full of blues, greens, reds, yellows, and purple. If it does, you are almost certainly eating foods that have a high phytonutrient index (PI) and are rich in the antioxidants you need to stay healthy and lose weight. This is the best way to fight rust and stay healthy over the long term, so make sure you are eating good carbs with a high PI that come from whole, unprocessed, fresh plant foods.

The basic Ultrametabolism food guidelines in chapter 16 describe a way of eating that eliminates the major dietary sources of oxidative stress and maximizes the intake of antioxidants.

Step 3: Use Herbs to Reduce Your Oxidation

The following herbal remedies can be a powerful addition to your antioxidation program. Try adding these herbs to your meals or taking them as supplements.

- Ginkgo
- Ginger
- Green tea polyphenols
- Pycnogenol or grape seed extract
- Milk thistle
- Rosemary
- Turmeric

Step 4: Use Supplements to Reduce Your Oxidation

Adding these supplements to your standard regimen can be a powerful aid in helping you reduce the effects of "rusting" so you can lose weight and be healthy.

A note on supplements: There are many special supplements that have multiple benefits across multiple keys. That is because the body runs on some very specific raw materials, and when things get rough (as in illness or obesity) more of these raw materials are needed to keep up with the demand. Just remember that some special supplements have multiple benefits and can help reduce inflammation and oxidative stress, improve mitochondrial function, and help you detoxify.

These supplements, suggested for inflammation (chapter 11), detoxification (chapter 16), and mitochondrial metabolism (chapter 13), are all excellent antioxidants.

- Reduced glutathione, the major antioxidant and detoxifier in our body
- *N*-acetylcysteine (NAC), an amino acid that boosts our own glutathione production

❖ Alpha-lipoic acid, a super antioxidant that helps reduce blood sugar and improve energy production in the mitochondria

❖ Coenzyme Q10, an antioxidant and key part of the energy production in the mitochondria

❖ NADH, part of the energy production cycle in the mitochondria

Step 5: Consider Testing for Oxidation in Your Body

The following tests are helpful for confirming high levels of oxidation and pinpointing the causes of rusting. If you scored in the moderate to high range on the quiz at the beginning of the chapter, I would strongly urge you to take these basic tests. You should also seek professional medical help if you scored high on the quiz.

Lipid Peroxides (TBARs) in Urine or Serum Test

This is a measure of rancid fats in your body and is linked to heart disease and many chronic illnesses.

Urinary 8-Hydroxy-2-deoxyguanosine Test

This is a urine test of damaged DNA from free radicals and one of the best ways to identify oxidative stress.

Assessment of Iron Overload Test

One of the most common inherited disorders makes us store too much iron, literally leading to accelerated rusting in the body. Any doctor can test your blood for transferrin saturation, ferritin, serum iron, and total iron-binding capacity.

Blood Tests for Antioxidant Levels

These can occasionally be helpful, including vitamin A, vitamin E, CoQ10, reduced glutathione, and beta-carotene.

Reduced Rust = Reduced Weight

By finding the causes of oxidative stress or rusting—which is primarily caused by empty calories and pollution—reducing them, and eating plenty of antioxidant-rich foods, as well as adding special herbs and supplements, you can go a long way to protecting yourself from the invasion of the free radicals, lose weight, and live longer.

The Stress of Processed Foods

Priscilla, 49, was under stress—oxidative stress. Struggling for years to lose weight, she had always tried the processed, packaged versions of diet plans. She used the TV-dinner, no-fuss method of food preparation. A box, a can, a package, and a microwave were all she needed. Busy, with a job, three kids, and a husband, and the victim of a traumatic childhood during which she had been subjected to overcooked, tasteless vegetables, and processed foods, she didn't know how to make real, whole food. This led to her progressive weight gain and frustrations with weight loss. Her belly grew larger, her cravings increased, and she developed more symptoms of sugar imbalance, including hypoglycemia, fatigue after meals, and, by the time I saw her, prediabetes.

Her blood tests showed rancid fat and damaged DNA, results of the overproduction of free radicals. These free radicals also damaged her ability to control her blood sugar and increased inflammation, contributing to her relentless difficulties losing weight. What she didn't realize was that by eating all those processed foods in boxes, packages, and cans—even if they were "weight loss" foods—she was missing one of the most important ingredients in her diet that promotes real weight loss: antioxidants, substances that come only in fresh, colorful packages, namely fruits and vegetables.

By learning how to prepare her own real food and giving up her familiar microwaved, overcooked, processed food, Priscilla was able to turn off the oxidative stress, which stopped the cycle of increased inflammation and sugar problems and helped her easily lose weight.

Summary

❖ Oxidation is a natural process, both within your body and everywhere in nature.

❖ When oxidation goes awry, damage is done to your cells that contributes to rapid aging, disease, and weight gain.

CHAPTER 13

TURN CALORIES INTO ENERGY:

Increasing Your Metabolic Power

Struggling for Oxygen: Boosting the Furnace

John had always struggled with his weight. At 50, the battle was fatiguing. He had been to every spa and tried every fad. His weight had gone up and down for years. He was otherwise remarkably healthy. He said he exercised, but it was not clear how much or how long or how consistent he was.

He was diligent about following the dietary suggestions I made and lost about 20 pounds using them. He had another 30 to go. So we put him on a special treadmill that measures your metabolic rate while exercising.

We measured the amount of oxygen he consumed and the amount of carbon dioxide he exhaled while exercising, which gave us an indirect measure of how active his metabolism was. Your ability to burn calories is dependent on the health, number, and efficiency of your mitochondria, the little powerhouses that produce energy in every cell. They take the food you eat, combine it with oxygen, and burn it; that is how you make energy. It is the core of your metabolism.

The amount of oxygen you're able to breathe per minute is directly tied to the number of calories you can burn per minute. When you breathe more oxygen, you burn more calories. That is why measuring John's oxygen consumption gave us a good indication of how efficient his metabolism was. When we put him on the metabolic treadmill, we found he consumed much less oxygen than he should have, based on his age, height, weight, and sex. His poor little mitochondria just weren't burning enough oxygen.

I sent him to an exercise physiologist, who provided a special exercise program called interval training. Interval training is a little like the wind sprints the gym teacher forced you to do in high school, where you ran very fast for a couple of minutes, then slowed down and repeated the process over and over.

This type of training helps the body become more effective at taking in oxygen, which in turn makes the muscle cells produce more mitochondria and makes the mitochondria you have more efficient. This helps you burn more calories, not only when you exercise but when you're at rest as well. It increases your metabolic fire.

After four months John returned, 30 pounds lighter. When we checked his metab-

olism again, he had increased his oxygen consumption (and his metabolism) by almost 50 percent.

Turning Up Your Metabolic Thermostat

When your house is cold, you simply walk over and turn up the thermostat. Wouldn't it be wonderful if when you gained a few pounds you could simply find the dial on your metabolic thermostat and turn it up? With the turn of a dial you could burn off the weight you gained. If only it were that easy.

New research reveals some interesting information about how we burn calories and why some of us store more than we burn. This new information can help you increase your metabolic power, burn more calories, and lose more weight (not to mention reverse the aging process).

So how can we avoid ending up old and fat? The answer lies in one word: *mitochondria,* the powerhouses of your cells. In the last chapter you learned that oxidative stress is primarily a by-product of burning calories and oxygen inside your mitochondria to create energy. About 5 percent of the oxygen we consume results in the production of free radicals. These damage the mitochondria, leading to impaired energy production, which causes a reduction in your metabolic power and your ability to burn calories.

But the good news is that you can reverse the damage that's been done to your mitochondria and turn up your metabolic fire. That's exactly what I'm going to teach you to do in this chapter. But before we get started on that, take the following self-assessment to see how efficiently your metabolic engine is running.

HOW POWERFUL IS YOUR METABOLIC ENGINE?

Score 1 point each time you answer "yes" to the following questions by putting a check mark in the box on the right. See page 78 for a reminder on interpretation of your score.

	YES
Do you experience chronic or prolonged fatigue?	☐
Do you have muscle-aching pain or discomfort?	☐
Do you have trouble falling asleep or staying asleep, or do you wake up early?	☐
Do you experience muscle weakness?	☐
Do you wake up tired despite a normal amount of sleep?	☐

Do you have poor exercise tolerance with severe fatigue afterward? ☐

Do you have trouble concentrating or memory problems? ☐

Are you often irritable and/or moody? ☐

Does fatigue prevent you from doing things you would like to do? ☐

Does fatigue interfere with your work, family, or social life? ☐

Have you been under prolonged stress? ☐

Did your symptoms of fatigue start after a severe stress of some sort, infection, or trauma? ☐

Have you been diagnosed with chronic fatigue syndrome or fibromyalgia? ☐

Do you have a history of chronic infections? ☐

Do you frequently overeat? ☐

Are you frequently exposed to environmental chemicals or heavy metals (pesticides, unfiltered water, food that is not organic, tuna, swordfish, or dental amalgams)? ☐

Have you been diagnosed with Gulf War syndrome? ☐

Have you been diagnosed with neurologic diseases such as Alzheimer's, Parkinson's, or ALS? ☐

If you don't want to write in this book, I've included this quiz in the companion guide that you can download by going to www.ultrametabolism.com/guide.

You can do something about how efficiently your mitochondria function. You can give yourself a metabolic tune-up. Now you just need to learn how.

Getting a Metabolic Tune-up

There is one thing that reduces stress on the mitochondria, significantly extends our life span, and causes us to lose weight. What is it? Calorie restriction![1]

But doesn't that go against the myth of the starvation syndrome from part I? Not exactly. Ideally, calorie restriction means eating enough calories to meet your daily metabolic needs, but no more. Diets that have proved effective in this regard are *restricted in calories* but *high in nutrients*. The rats being tested had their nutritional needs met while eating only enough to meet their most essential metabolic requirements.

In any event, it doesn't sound like a fun proposition. You will be thinner and live longer, but you and everyone around you will be miserable because you will be hungry all the time. Is there an alternative? Yes. It is the best-kept secret of the recent past, one that no one is talking about. There is a way to give yourself a "metabolic tune-up" without starving yourself.

To understand how you can increase your metabolic power, you first need to understand how the mitochondria work and how they burn calories. Then you must learn what can go wrong and slow down your metabolism. Once you have identified the causes of mitochondrial problems, you can fix them and learn how to give yourself a metabolic tune-up. Then you can burn calories more easily and more efficiently, making it much easier to lose weight.

How Mitochondria Work and How They Determine Your Metabolic Rate

Mitochondria are the parts of your cells that combine the calories you consume with oxygen and turn this mixture into energy, used to run everything in your body. A single cell may have anywhere from 200 to 2,000 or more mitochondria. Cells that work hard, such as those in the heart, liver, or muscle, contain the greatest number. You wouldn't be able to breathe, much less walk out your front door, if it weren't for these little powerhouses. They are responsible for keeping you alive.

The rate at which your mitochondria transform food and oxygen into energy is called your metabolic rate, and it is determined by two factors: the number of mitochondria you have and how efficiently they burn oxygen and calories. The more mitochondria you have and the more efficiently they consume oxygen, the faster your metabolic rate, the easier it is for your body to burn calories, and the more energy you have.

Fortunately, you have the ability to dramatically influence both of these factors. The answer lies in one word: exercise.

Get Moving

I know you were hoping that I would avoid the subject, but it is as inevitable as death and taxes. Besides eating breakfast, exercise is the *only* thing that has been correlated with long-term, sustained weight loss. This is for one primary reason: the best way to increase the number of mitochondria in your body and improve the functioning of the ones that are already there is to exercise.

When you exercise, you increase your muscle mass and increase your oxygen intake. These are both important factors in positively affecting your mitochondria. By increasing your muscle mass, you increase the number of cells in your body that contain large numbers of mitochondria (remember that your muscles have one of the highest concentrations of mitochondria).

When you increase your oxygen intake, you direct your mitochondria to process more oxygen more quickly. When you work out, they work out. That means they get better and better at consuming oxygen. In short, exercise increases your metabolic power.

This has important ramifications that go well beyond the number of calories you burn while you exercise. What most people don't realize is that these calories are only part of the story. *The calories you burn when not exercising are just as important.* By increasing the number and function of the mitochondria in your body, you increase your ability to burn calories at rest; hence you increase the amount of calories burned even while you sit at your computer checking e-mail or sleep.

Why Yo-Yo Dieters Can't Lose Weight

Loss of muscle is actually one of the reasons yo-yo dieters can't seem to lose weight. They lose and gain over and over again. This in itself is annoying, but the big problem is that when they lose weight, they lose half muscle and half fat (the body holds on to fat preferentially as a natural survival mechanism), and when they gain the weight back, they gain back all fat for various reasons. Since muscle is much more metabolically active and burns seventy times as many calories as fat cells, losing muscle slows down their metabolism, making it easier to gain weight and harder to lose it, even while they eat the same number of calories.

Unfortunately, it's easier to gain and retain fat than it is muscle, and while losing fat takes work, losing muscle is (literally) effortless. Loss of muscle happens after the age of about 35 if you do nothing to prevent it (smaller losses can occur even earlier if a person is sedentary). This loss of muscle has a tremendous impact on your ability to lose weight. This is why older people have such a tough time losing weight.

Sarcopenia = Muscle Loss—The Common Link to Aging and Obesity

A special machine called a DEXA provides a body composition test that measures the amounts of muscle and fat in people. It often shows muscle loss, known as *sarcopenia* (sarco = muscle and penia = loss). Although some people are obviously overweight and you expect them to have a higher level of fat, others are what we call "skinny fat people" because they look skinny but have very little muscle.

It's the *ratio* of fat to muscle that makes the biggest impact on health and metabolic balance, and a person's outward appearance can be deceiving in this regard.

Suppose you lose 10 ounces of muscle a year. Even if you gain 10 ounces of fat per year during this time, the numbers on the scale may not change significantly and your clothes may not fit differently. In other words, at age 70 you could end up the same *weight* you were when you were age 20 but be twice as *fat* because your muscle tissue was replaced with fat! We call this "metabolic obesity" or "skinny fat syndrome." It has the same dangerous consequences for your health as being obese. As you lose muscle, you lose mitochondria and your metabolism slows.

> More muscle = more mitochondria, which is central to boosting your metabolic power.

The Fat Genes:
Slow Mitochondria Cause Diabetes and Weight Gain

It is clear that obese people have fewer and more poorly functioning mitochondria than thin people do. Perhaps that's why thin people stay thin and obese people stay obese. But then again, maybe not! Recent scientific discoveries tell us that thin relatives of people with type 2 diabetes have smaller and more slowly burning mitochondria than other thin people do. This surprising finding has turned our perceptions and assumptions upside down.[2] It tells us that those who have genetically smaller or poorer-functioning mitochondria are more likely to gain weight and get diabetes.

Diabetes and mitochondrial function are linked, and some scientists say that the primary cause of diabetes is damage to the mitochondria.[3] Too many calories, too much sugar, high-fructose corn syrup (HFCS), high-glycemic-load or rapidly absorbed carbohydrates, and too many saturated and trans fats all cause insulin resistance, damage the mitochondria, and lead to type 2 diabetes.

So anything that interferes with your system—oxidants and free radicals, sugar, trans fats, inflammation, and the vicious cycle of counterproductive hormonal messages coming from the fat cells themselves (TNFα, resistin, IL-6)—puts you into a tailspin of impaired metabolism and weight gain.

The take-home message is that even though you might have the genes that predispose you to insulin resistance and obesity, you can turn off the activity of those genes through exercise. You can also increase your mitochondrial function and get a metabolic tune-up by adopting certain key nutritional principles and taking special supplements that help turn off slow-metabolism genes and turn on the metabolism-boosting genes.

Whether or not you are genetically predisposed to gaining weight, you can turn up your metabolic power by following the recommendations given in this chapter. Follow these five steps to customize the UltraMetabolism Prescription for your own particular body's needs:

- ❖ **Step 1:** Eliminate the causes of mitochondrial damage.
- ❖ **Step 2:** Exercise intelligently.
- ❖ **Step 3:** Eat foods that turn up your metabolism, and avoid foods that turn it down.
- ❖ **Step 4:** Use supplements that give you a metabolic tune-up.
- ❖ **Step 5:** Test your mitochondria.

If you are having problems with this UltraMetabolism key and you don't have the metabolic power you need, you can reverse the effects of your damaged mitochondria by following these steps. By doing this, you can customize the UltraMetabolism Prescription for your own particular needs. This allows you to turn on genes that cause you to lose weight and turn off the ones that have been making you gain weight. Let's look at exactly how you can make this five-step process work for you.

Step 1: Eliminate the Causes of Mitochondrial Damage

To increase your metabolic power, it is critical to remove or reduce those things that damage your mitochondria and to use foods and supplements that help improve their function and provide a metabolic tune-up. As mentioned in chapter 12, the major factor that damages the mitochondria is oxidative stress, or free radicals. Too many calories and not enough nutrients (and phytonutrients) in your diet will increase the damage. The solution: eating a nutrient-dense, phytonutrient-rich, unrefined, unprocessed, real, whole-food diet.

This way of eating is at the heart of the UltraMetabolism Prescription. Following the eating program in part III will help you to keep your mitochondria healthy.

You also need to identify other causes of damage to your mitochondria. These might include thyroid hormone problems (we will address this in chapter 14); toxins that damage mitochondria, such as mercury (see chapter 15); chronic infections (see chapter 11); anything that causes inflammation;

and taking certain medications (such as statins, which deplete coenzyme Q10, a nutrient important for energy metabolism). If you make an effort to balance these elements on your own using the program in this book and still seem to have problems, you might want to seek a doctor's help for advanced testing in the areas that have potential to damage the mitochondria. See the Resources or www.ultrametabolism.com for recommendations for finding physicians.

Step 2: Exercise Intelligently

We were genetically designed to move, to use our bodies. That's how our bodies flourish. If you want to heat up your metabolic fire, you need to exercise. If you currently do no exercise, begin to do so. If you already exercise, do more.

I must confess: I hate exercise. You will almost never catch me in a gym. But I love to *play*. I play tennis and basketball, I ski, bike through the Berkshire hills, jump on the trampoline, hike up mountains with friends, dance to R & B, wrestle with my children, and run with my dog and my iPod. But exercise—never.

The key is finding something you love to do, or a hundred things you love to do, and doing them. Our ancestors never exercised, but their life was filled with movement. Your genes depend on movement to produce messages that lead to a healthy metabolism. The more vigorous the activity, the better, but just getting rid of your TV remote control and getting up off the couch every time you want to change the channel can lead to significant weight loss over the course of time.

Research on exercise has found that thirty minutes of aerobic exercise five times a week helps you achieve most health benefits but that doing up to sixty minutes five times a week is needed for weight loss.

Not to fear: there are ways to exercise smarter, not longer, and achieve more benefits, and I am going to show you how to do that. Eventually you will want to combine an aerobic exercise routine incorporating the science of interval training with a good strength-training regimen. But before we get started on that, I'd like to show you how simply moving more in your daily life can add up to significant weight loss benefits and remarkable long-term weight loss.

Move a Little Every Day

A study in the prestigious journal *Science* with the technical title "Interindividual Variation in Posture Allocation,"[4] otherwise known as fidgeting, reported that simply getting up from your seat more often may burn as much as an additional 350 calories per day. That can amount to a weight loss of 36 pounds in a year.

In the study, ten lean and ten obese volunteers were studied using sophisticated motion sensors developed by NASA. Their movements were recorded every half second for ten days. The overweight individuals moved, on average, two hours less per day. They were missing a very important factor in weight loss, nonexercise activity thermogenesis, or NEAT—they weren't fidgeting enough.

The government's Small Steps campaign (www.smallstep.gov) provides numerous ways to increase your daily activities. Fidget, get up to change the TV channel, or better yet, leave the TV off and clean up the basement! A simple way to track your daily physical activity is by using a step counter. Clip it on your belt, and it counts how many steps you take. Try to get up to 10,000 steps a day. Remember to adjust your "posture allocation" and lose weight!

Here are some simple steps recommended by the U.S. Department of Health and Human Services. Pick five of them and try them this week.

- Do sit-ups (or any form of activity) in front of the TV.
- Walk during your lunch hour.
- Walk instead of driving whenever you can.
- Take a family walk after dinner.
- Walk to your place of worship instead of driving.
- Get a dog and walk it.
- Join an exercise group.
- Do yard work.
- Get off the train or bus a stop early and walk.
- Work around the house.
- Bicycle to the store instead of driving.
- Go for a half-hour walk instead of watching TV.
- Wash your car by hand.
- Pace the sidelines at kids' athletic games.
- Park further from the store and walk.
- Ask a friend to exercise with you.
- Exercise indoors with a video if the weather is bad.
- Take a walk instead of a cigarette or coffee break.
- Play with your kids thirty minutes a day.
- Dance.
- Walk briskly in the mall.
- Explore new physical activities.

❖ Vary your activities, for interest and to broaden the range of benefits.

❖ Take stairs instead of the escalator.

❖ Walk to a coworker's desk instead of sending an e-mail.

❖ Use a snow shovel instead of a snowblower.

❖ When walking, go up the hills instead of around them.

❖ Buy a set of hand weights and play a round of "Simon says" with your kids: you do it with the weights, they do it without.

❖ Reward and acknowledge your efforts.

Breathe More to Build Up Your Metabolic Engine: Aerobic Training

Aerobic training is an important part of your exercise regimen. It increases the oxygen supply to your body, making your mitochondria more efficient at consuming oxygen. This boosts your metabolic engine. Anything that speeds up your respiration and your heart rate counts as aerobic exercise. Walking, jogging, biking, swimming, and playing tennis are just a few examples.

Ideally, you want to do some type of aerobic training for thirty to sixty minutes five days a week. In addition, you want to raise your pulse to 70 to 85 percent of your maximum heart rate. This is easily calculated by subtracting your age from 220 and then multiplying the resulting amount by .70 to .85 to get your target heart rate range. For example, right now I am 45 years old, so if I subtract 45 from 220, which is 175, and then multiply that by .70 and .85, I find that my target heart rate range is 122 to 148.

This is your target heart rate each time you exercise. Reaching it will provide you with maximum health benefits and dramatically increase your ability to lose weight. Exceeding your target heart rate is not a bad thing. In fact, it is *essential* during interval training; pushing yourself harder for short bursts has remarkable effects. It is a way of exercising *smarter* and getting more benefit out of the time you do exercise.

If you haven't been exercising much, you will want to start slowly and build up over time. I would recommend starting with no more than ten minutes of low-impact aerobic conditioning daily if you have been leading a sedentary lifestyle. This can be as simple as walking around the block. Over time you can increase your efforts. For example, if you start with ten minutes a day one week, you can increase it to fifteen or twenty minutes a day the following week. Continue that pattern until you are at the optimum daily level.

It helps to vary your routine so you don't get bored (this is one of the advantages of incorporating interval training in your program; see page 168).

You want to give yourself all the positive reinforcement you can, and boring yourself doesn't really help.

To this end you might consider training with a friend, listening to music while you exercise, or engaging the help of a private trainer. Anything you can do to motivate yourself to exercise is worth it. If you exercise consistently, you will feel lighter, look better, and have more energy.

I would also strongly encourage you to incorporate interval training into your routine every other day. Interval training is a particular style of aerobic exercise that has been shown to have incredible effects on your ability to lose weight.

Burn More Fat While You Sleep: The Science of Interval Training

Everyone knows that you burn calories when you exercise, which promotes weight loss. But is there a way to burn more calories *after* you exercise, during rest or sleep? Is there a way to *exercise less* and *get more benefits?*

The answer is yes. The key is something called *interval training,* short bursts of high-intensity exercise followed by longer periods of lighter exercise—what we called wind sprints in high school or what the Swedes call *fartlek,* or "speed play." Though interval training was designed to help professional athletes maximize their performance, the average person can greatly benefit from it as well.

Dr. Martin Gibala, Ph.D., from McMaster University in Canada, has shown that you can exercise smarter, not longer, and achieve more benefits. This can involve a few short bursts of all-out activity (about 90 percent of your capacity; 100 percent is the exertion you would spend running from a tiger) for thirty to sixty seconds with a three-minute rest period between each burst at about 50 to 60 percent of your maximum capacity. Your maximum capacity is 220 minus your age unless you are on a beta-blocker medication. This should total about twenty to thirty minutes of interval training two or three days a week. It can dramatically improve your fitness level, leading to more fat loss than traditional endurance exercise programs do.

When a group of Canadian researchers from Laval University[5] compared regular endurance or aerobic conditioning to interval training, they discovered something remarkable. The first group (endurance training) worked out for a longer time period (twenty weeks compared to the interval group's fifteen), exercised longer for each training period (forty-five minutes compared to thirty), conducted more total workouts (ninety sessions compared to sixty), and burned twice as many calories during individual exercise periods (120 millijoules versus 59).

By the simple laws of physics, the endurance group should have lost more fat than the interval group. But by exercising only half as much, the interval group reduced their body fat *nine* times more than the endurance group. The reason? They increased their resting metabolic power. Interval training allowed them to burn more calories at rest than the endurance-training group did.

Could this be true? Isn't it calories in, calories out that control our loss of body fat? Again we see that this is a myth. Science tells us a number of important things about the benefits of including interval training in our exercise routine:

1. We improve our fitness level, our ability to utilize oxygen. And the more oxygen we use, the more calories we burn.
2. We increase *postexercise* fat burning and calorie expenditure, even during rest or sleep.
3. We can exercise for less time and achieve greater fitness and weight loss benefits.
4. We can naturally increase our growth hormone levels, which promotes fat burning and muscle building; not a bad bang for the buck!

So what are the pros and cons of interval training? On the positive side, very intense exercise is time-efficient and more effective for weight loss than regular aerobic exercise is. You need to do it only two to three days a week, and you can do it for shorter periods of time. It has the same health benefits as regular aerobic exercise, and maybe more. It leads to greater fat loss. Plus, varying the intensity of exercise (something that is built into every interval program) makes your routine more engaging and less boring.

On the downside, you experience discomfort. Pushing yourself to 80 to 90 percent of your maximal capacity makes you short of breath and causes fatigue in your leg muscles. If you are over 30 years old, you should have a complete physical exam before starting an interval training program. If you are completely sedentary, you will need to engage in a more gentle exercise routine for four to twelve weeks before beginning interval training. Warming up before intervals is also critical, to prevent muscle strain or injury.

In the end, the tremendous benefits far outweigh a few seconds of discomfort here and there. I would recommend instituting the following interval program as soon as possible.

Interval Training Guidelines

What follows is a step-by-step overview of how interval training works. There is a version for people who are just beginning to exercise as well as one for those who are a little more advanced and have already been working out regularly. If you don't fit into either of these categories (i.e., if you can't walk for thirty minutes at 3.5 miles per hour), you should build up your aerobic exercise program before you start incorporating interval training.

Beginner (for someone who can walk for 30 minutes at 3.5 miles per hour):

1. Warm-up: 5 minutes of walking at 3.5 miles per hour.
2. Speed up and walk at 4.0 miles per hour for 60 seconds.
3. Slow down and stroll at 3.0 miles per hour for 75 seconds.
4. Repeat steps 2 and 3 five more times.
5. Finish with 5 minutes of walking at a comfortable pace to cool down.

Advanced Interval Training Program

1. Warm-up: 5 minutes of jogging or cycling at the lowest possible percentage of your all-out effort.
2. Run or cycle for 60 seconds at about 80 to 90 percent of your all-out effort. Your leg muscles should fatigue in about 1 minute. (Basically, the speed you'd run or cycle at to save your life equals 100 percent of your all-out effort. From there, adjust how fast and hard you work so your output reflects the recommended percentage.)
3. Slow down to 50 percent of your all-out effort for 75 seconds. (Make sure you slow down to this very light pace.)
4. Repeat steps 2 and 3 five more times.
5. Finish with five minutes at 30 percent of your all-out effort to cool down.

Muscle Your Way to Weight Loss: Strength Training

With interval training you can make your cells smarter, improve the efficiency of your mitochondria, and even create a few more. But you also need to do something to stop the inevitable loss of muscle that happens with aging. One exercise I have my patients perform in the office is to stand up out of a chair without leaning forward or using their arms. It is incredible how many people (even younger people) have lost so much muscle that their thighs can't even lift their body weight off the chair without some help. Try it yourself now.

Strength training helps to increase muscle size and strength, increases the number of mitochondria in your body, and can boost your metabolic rate so you burn more calories at rest or sleeping. Find something you like, vary it, but try *something*. Using your own body weight by doing an activity such as yoga (now practiced by 17 million Americans), stair climbing, push-ups, or squats can be great. Finding a gym and using weights is another way to build muscle. If you have never lifted weights, be sure to get advice from the fitness trainer in the gym on using proper technique and form to avoid injury.

Ideally, you want to build up to two sets of eight to ten repetitions of a weight that leads to muscle fatigue for each major muscle group. A twenty-minute routine two to three times a week can cover all the bases. Who doesn't have forty to sixty minutes a week to invest in health and metabolism boosting?

All of the things in this book are additive and work together. I have had many patients lose significant amounts of weight without exercise. But at some point they get stuck, can't seem to lose that last 10 to 15 pounds, and need something to get them off that plateau—and aerobic training, including intervals and strength training, can take people all the way.

It also reduces your appetite and sets into motion all the hormone-, brain-, and immune-balancing chemicals that promote a healthy weight and metabolism. In fact, exercise is one of the most potent anti-inflammatories and antioxidants known (not to mention that it helps control hunger signals, improves insulin sensitivity, burns off stress chemicals, boosts thyroid function, and helps your liver detoxify).

Step 3: Eat Foods That Turn Up Your Metabolism, and Avoid Foods That Turn It Down

Some foods help turn up your metabolism; others harm your mitochondria and turn it down. By following the food guidelines in the UltraMetabolism Prescription, you will know which foods to eat and which to avoid so you can burn fat and stay healthy.

Step 4: Use Supplements That Give You a Metabolic Tune-up

Adding the right supplements to your daily regimen can help give you a metabolic tune-up. Try these supplements to boost your mitochondrial horsepower:

- **N-acetylcysteine (or NAC):** Helps restore glutathione in the cells, the most powerful antioxidant in the body
- **Acetyl-L-carnitine:** Transports fat into the mitochondria
- **Alpha-lipoic acid:** Helps protect the mitochondria from oxidation
- **Coenzyme Q10:** Helps increase energy production
- **NADH:** Helps increase energy production
- **Creatine powder:** Helps provide energy for the muscle cells and metabolism
- **Amino acids (arginine and aspartic acid):** Raw materials used in cellular metabolism needed for the mitochondria
- **D-ribose:** The raw material for energy production and the creation of ATP in the cells

Step 5: Test Your Mitochondria

Testing for mitochondria is fairly specialized. In my practice I have used two tools that are not in common use but are increasingly available. The first is a cardiometabolic stress test. The other is a urinary organic acid profile. There are also more sophisticated ways to test your mitochondria, if needed, that can be done by muscle specialists. These tests are not critical to fixing your mitochondria, but I use them in my practice to help me refine my recommendations.

Cardiometabolic Stress Test

This test measures the rate at which you consume and burn oxygen, or your VO_2 max. It is very simple: the more oxygen you burn, the more calories you burn. VO_2 max is a measure of the oxygen you consume during exercise, which is directly related to your fitness level and calorie-burning capacity. It must be performed by a physician with an exercise physiology or sports medicine practice. A modified form of this test is often available at higher-end fitness centers.

Organic Acids Test

Organic acids is a urine test that measures the by-products of metabolism, nutritional deficiencies, and more. While the cardiometabolic test looks indirectly at the number and power of the mitochondria, the organic acids look at their biochemical efficiency. Only a few specialized labs (see www.ultrametabolism.com/tests) measure these, and they are challenging to interpret, but they often provide remarkable clues to metabolic and nutritional problems.

The UltraMetabolism Prescription and Your Metabolic Engine

Optimizing your metabolic engine is an important way to increase your ability to lose weight and is often the solution to losing those last few pounds. This is done through exercising, eating as outlined in the Ultra-Metabolism Prescription, eliminating factors that damage your mitochondria, and adding supplements to your daily regimen. The supplements can help protect your mitochondria and balance your hormones and blood sugar. Understanding the role of mitochondria in weight loss and aging and learning to fix problems is one of the most exciting new areas of research in medicine.

The Mitochondria/Diabetes Connection: Fixing the Damaged Metabolic Powerhouse

Diane was in her late 60s when she came to see me. A widow, she lived alone and was estranged from her family but had a few close friends. She had had diabetes for ten years, as well as rheumatoid arthritis, allergies, hypothyroidism, reflux, and angina. And she was always tired and depressed, though it was not hard to imagine why.

She weighed more than 250 pounds and was five feet four inches tall and had gained and lost weight many times: the last time, she had lost 100 pounds but hadn't kept it off very long. She was on a cocktail of the "best" pills to treat her ills: a diabetes pill, an antidepressant, a cholesterol-lowering pill, pills for high blood pressure, allergies, arthritis, and thyroid. She was even on estrogen for menopause. No symptom was left untreated. But she still felt bad.

She soothed herself with a diet full of bread, pasta, muffins, and ice cream and was too tired to exercise. Her blood tests weren't much better than her diet. Blood tests revealed that her blood sugar and insulin levels were high despite the medication, her triglycerides were elevated, and her liver was fatty. She also had a very high level of inflammation, with a C-reactive protein test of 10 mg/L (normal is less than 1).

Many diabetics have a problem burning the calories they eat, which promotes fat accumulation in their cells for two reasons. The first may be the presence in their diet of rapidly absorbed carbohydrates, such as white bread and muffins. The second is that the "production line" may be jammed. The little factories (the mitochondria) that burn food slow down and the fat in line for combustion accumulates. There may be some genetic predisposition to this problem, but once identified it can be overcome with treatment. Special tests look at the function of the mitochondria by evaluating organic acids in the urine; they identify which steps of the fat-burning process may be moving too slowly.

We tested Diane and found she had many problems processing fats and carbohydrates through her mitochondria. Though she had tried many treatments for her weight and diabetes, none had been helpful in the long run. By addressing the damaged mitochondria through reducing the sugars in her diet, exercising moderately, and taking some special supplements, including lipoic acid, CoQ 10, and carnitine,[6] she was able to get her mitochondria burning all that fat. She was also able to stop taking a number of medications.

A year later she had lost 45 pounds, her inflammation was gone, her liver was no longer fatty, her blood sugar and triglycerides were normal, and she no longer had diabetes. More important, she had her life back, with energy and joy back in her step. When we tested her fat, carbohydrate, and energy metabolism, it had all returned to normal.

Summary

- Your mitochondria are the parts of your cell that convert calories and oxygen into energy. The more you have and the more efficient they are, the more calories you can burn both when exercising and at rest.
- One of the reasons yo-yo dieting doesn't work is because you lose muscle and replace it with fat. When this happens, you cut your metabolic power in half.

CHAPTER 14

FORTIFY YOUR THYROID:

Maximizing the Major Metabolism Hormone

Depression and Obesity: Missing the Obvious

*M*elissa was a single 38-year-old Wall Street investor working crazy hours. After September 11, 2001, like many people, she felt depressed, and her doctor started her on an antidepressant. She gained 40 pounds. She ate fairly well, had an egg in the morning, and didn't eat sweets, although she craved carbohydrates.

Melissa exercised regularly, had many friends, and enjoyed life. However, she was tired and constipated, complained that her periods were irregular, and suffered from premenstrual syndrome (PMS) with bloating and irritability before her periods. Her doctor treated this with a birth control pill. Her cholesterol was terrible at 275 mg/dl, and she had inflammation and a high insulin level.

Let's see: fatigue, constipation, depression, trouble losing weight, weight gain, PMS, high cholesterol, insulin, and blood sugar levels, and inflammation. How are they related? None of the doctors she saw had looked in the right place: at her thyroid. Her regular thyroid test or TSH was "normal," but she had antibodies acting against her thyroid gland (her body was essentially treating it as a foreign invader), and the active hormone, the most important one, T3, was very low.

A few months after starting on Armour thyroid (a combination of all thyroid hormones, including the active T3 and inactive T4), she lost 30 pounds and all her health problems evaporated.

A Whole-Listic Doctor:
Connecting the Whole List of Complaints

Because doctors are trained to look for "real disease" (with "textbook" symptoms that can be diagnosed and matched to a prescription drug), subtler, chronic symptoms are often ignored. But those symptoms are the clues to the deeper mystery of what is wrong with you. Many of you have likely gone to the doctor, complaining of various symptoms such as fatigue, depression, muscle cramps, menstrual problems, difficulty losing weight, constipation, memory trouble, or joint pain. Maybe you were prescribed an antidepressant, told you were just getting older, or, worse, to eat less and exercise more.

I call myself a "whole-listic" doctor because I take care of people with a "whole list" of problems. And the patients with the biggest lists are those whose thyroid systems are not functioning well. Many of you may suffer from some or all of the symptoms listed above, and they are all clues to a low-functioning thyroid.

The thyroid system plays a critical role in your metabolism. Along with insulin and cortisol, your thyroid hormone is one of the big three hormones that control your metabolism and weight. Twenty percent of all women (and about 10 percent of men) have a sluggish thyroid, which slows their metabolism, and half of them are undiagnosed. To make matters worse, many of those who *are* diagnosed are not treated optimally.

In addition to understanding whether you are being affected by the symptoms listed above, taking the following test will give you an idea of whether or not you have problems with your thyroid.

ARE THYROID PROBLEMS CONTRIBUTING TO YOUR STRUGGLE WITH WEIGHT?

Score 1 point for each time you answer "yes" to the following questions by putting a check mark in the box on the right. See page 78 for a reminder on interpretation of your score.

	YES
Are your skin and fingernails thick?	☐
Do you have dry skin?	☐
Do you have a hoarse voice?	☐
Do you have thinning hair, hair loss, or coarse hair?	☐
Are you cold when everyone else is warm?	☐
Do you have cold hands and feet?	☐
Is your basal body temperature less than 97.8 first thing in the morning (underarm basal body thermometers are available at most drugstores)?	☐
Do you have muscle fatigue, pain, or weakness?	☐
Do you have heavy menstrual bleeding, worsening of premenstrual syndrome, other menstrual problems, and/or infertility?	☐
Have you experienced a loss of sex drive (decreased libido)?	☐

(continued)

Do you have severe menopausal symptoms (such as hot flashes and mood swings)? ☐

Have you experienced fluid retention (swelling of hands and feet)? ☐

Do you experience fatigue? ☐

Do you have low blood pressure and heart rate? ☐

Do you have elevated cholesterol? ☐

Do you have trouble with memory and concentration or "brain fog"? ☐

Do you wake up tired and have trouble getting out of bed in the morning? ☐

Do you have a loss or thinning of the outer third of your eyebrow? ☐

Do you have trouble losing weight, or have you recently experienced weight gain? ☐

Do you experience depression and apathy or anxiety? ☐

Do you experience constipation? ☐

Have you been diagnosed with autoimmune disease (e.g., celiac disease, rheumatoid arthritis, multiple sclerosis, lupus), allergies, or yeast overgrowth, all of which can affect thyroid function? ☐

Are you or have you been exposed to radiation treatments? ☐

Are you currently or have you been exposed to environmental toxins? ☐

Do you have a family history of thyroid problems? ☐

Do you drink chlorinated or fluoridated water? ☐

If you don't want to write in this book, I've included this quiz in the companion guide that you can download by going to www.ultrametabolism.com/guide.

Thyroid dysfunction is often a problem that requires additional testing and a doctor's assistance to overcome it. Being informed about what tests to take is helpful. In addition, there are some things you can do to help heal your thyroid. This chapter will give you all those tools.

Diagnosing Thyroid Problems: How Doctors Miss Subtle Clues

When I studied medicine, I thought that diagnosing and treating low thyroid function was straightforward and simple. A patient shows up with the classic symptoms: tired, cold, swollen, overweight, dry skin, soft nails, hair loss, constipation, muscle cramps, impaired memory, and depression. Her blood tests for thyroid are abnormal. You give her Synthroid, the standard treatment for thyroid disease, which is *only* the inactive thyroid hormone called T4. You fix the problem, and everyone goes away happy. I was wrong.

After twenty years of medical practice, I now think that diagnosing and treating thyroid problems is one of the most complicated and difficult things I do—and when I get it right, it has the most profound effect on a person's overall health, weight, and metabolism of any of the treatments I prescribe.

Thyroid treatments are also one of the most controversial areas of medicine today. While the protocol for overt thyroid dysfunction is fairly well defined, doctors don't agree on the best way to diagnose and treat people with subtle thyroid problems. The guidelines change frequently, and there is a delay in getting the information into medical practice.

This is partly because the research in this area is controversial and not well understood. Some of what may be the most effective treatments have not yet become part of mainstream medical practice. What I have learned comes from studying for hundreds of hours, doing thousands of tests for thyroid problems, and treating thousands of patients for this problem.

The Function of the Thyroid Gland and the Importance of Nutrition

The thyroid gland is a small endocrine gland in your neck that makes two major thyroid hormones: about 93 percent is T4, the inactive form, and about 7 percent is T3, the active form. The T4 made in the thyroid gland is then converted into T3 in the liver.

Many dietary factors, as well as lifestyle and environmental factors, affect this process. The thyroid is part of your endocrine or hormonal system. The main role of thyroid hormone is to stimulate metabolism, and it affects almost every function of the body. That's why it can cause so many different symptoms. Thyroid hormone interacts, or has "cross talk," with all the other hormones in your body, including insulin, cortisol, and your sex hormones.

The production and release of thyroid hormones in the thyroid gland are regulated by a feedback system in your brain—the hypothalamus and pituitary glands—which make thyroid-releasing hormone (TRH) and thyroid-stimulating hormone (TSH), respectively. If everything works as designed, you will make what you need, and the T4 will be converted to T3.

The T3 then acts on special receptors (such as the PPAR family we have talked about) on the nucleus of the cell that send a message to your DNA to turn up your metabolism, to increase the fat burning in your mitochondria, and generally to make every system in your body work at the right speed. This is why T3 lowers your cholesterol, improves your memory, keeps you thin, promotes regrowth in cases of hair loss, relieves muscle aches and constipation, and even cures infertility in some patients.

If you produce too little T3, or the T4 you produce is not properly converted into this active thyroid hormone, your whole system goes haywire. Your metabolism and mitochondria don't get the proper signals, you gain weight, and you suffer from the symptoms described earlier. In addition, you can become more inflamed, develop additional problems with your insulin levels, and have a more difficult time metabolizing sugar in your blood, all of which further compromise your health and your ability to lose weight. One study showed that subclinical hypothyroidism (see below) causes both high levels of C-reactive protein and elevated insulin levels,[1] further indicators that the stability of your thyroid can have a dramatic impact on your health.

This wouldn't be such a concern if thyroid disorders could be readily diagnosed and treated. The problem is, they can't. Hypothyroidism, the name for the production of too little thyroid hormone, is a vastly underdiagnosed health problem in this country.

Hypothyroidism: An Undiagnosed Epidemic

Why is it so difficult to diagnose and treat low thyroid function? The main reason is that the symptoms are not very specific and are often present for many reasons besides thyroid disorders.

Anyone can diagnose a heart attack if he sees someone who is pale and sweaty and clutching his chest, complaining of crushing pain in the chest and down the left arm. Thyroid problems are completely different. Even if you have all the symptoms of low thyroid function, they may still easily be ignored. You may not even realize that the problem is with your thyroid gland.

Even if you have the foresight to go to the doctor, your doctor may use typical tests to test for thyroid problems and find that your thyroid *appears* to be functioning in the normal range. But many times doctors don't do the right tests or don't do enough tests; hence your thyroid problems go undetected. You may be told you have borderline thyroid problems or subclinical thyroid disease and your doctor will watch it. What will he or she watch for? For you to get really sick?

Thyroid problems are actually extremely common. More than 10 percent of the overall population and 20 percent of women over 60 years of age have subclinical hypothyroidism. "Subclinical" implies no symptoms and slightly abnormal thyroid tests. What it really means is subtle symptoms that are often missed by doctors!

Even people who have "normal" thyroid results but suffer from symptoms may benefit from thyroid treatment. It just depends on how you define "normal." If you are a seven-foot-tall basketball player, it might be normal to be 300 pounds, but not if you are a five-foot, three-inch female. If you were a Martian landing in America in the twenty-first century, you might think that it is normal to be overweight because more than 60 percent of our population is overweight. But that doesn't make it normal!

The normal values in medicine are constantly going down as we recognize that what we thought was normal isn't. In 1998, normal weight was a body mass index (BMI; kilograms per meter squared) of 27; now it's 25. Prior to 2001, normal cholesterol was 240 mg/dl; now it's 200. Normal LDL was 140 mg/dl; now it's 100. Normal blood sugar was once 140 mg/dl; now it's less than 100. Normal blood pressure was 140/90 mm Hg; as of August 2004 it was 115/75 (based on the *Seventh Report of the Joint National Committee on Prevention, Detection, Evaluation and Treatment of High Blood Pressure*).

Why does this happen? We are simply wising up and recognizing that more subtle changes in function can have significant health consequences. The same is true with thyroid disease, but mainstream medicine has not yet caught on. We doctors need to rethink how we approach thyroid problems by:

1. Recognizing the problem through analyzing a patient's medical history
2. Using the right tests
3. Properly diagnosing and treating the causes of thyroid dysfunction
4. Supporting the thyroid function with lifestyle changes, diet, and supplements
5. Prescribing thyroid hormone preparations and dosages specifically designed for individual patients

Right now, these things are not being done by the medical community at large.

However, there are a few things you can do to improve the function of your thyroid, and armed with the proper information you have a much better chance of being diagnosed correctly. In the remainder of this chapter I'm going to teach you what you can do to optimize your thyroid and tell you what tests are available, which ones you need to take, and what hormone treatment therapies are available.

Fixing the Thyroid: An Integrative Approach

Correcting thyroid problems is crucial to having a healthy metabolism. It involves more than simply taking a thyroid pill. It involves nutritional support, exercise, stress reduction, supplements, reducing inflammation, and sometimes eliminating certain foods and detoxification from heavy metals (such as mercury and lead) and petrochemical toxins (such as pesticides and PCBs).

To integrate all of these elements and create a successful set of techniques to cope with your thyroid problems, you can follow the six-step model laid out in this chapter:

- **Step 1:** Eliminate the causes of thyroid problems.
- **Step 2:** Exercise and take saunas.
- **Step 3:** Eat foods that provide nutritional support for your thyroid, and avoid those that don't.
- **Step 4:** Use supplements that support your thyroid.
- **Step 5:** Have your thyroid tested.
- **Step 6:** Choose the right thyroid hormone replacement.

If you are having problems with this UltraMetabolism key and your thyroid isn't functioning as well as it should, you can reverse the effects of unbalanced thyroid by following these steps. By doing so, you can customize the UltraMetabolism Prescription for your own particular needs. This will allow you to turn on genes that cause you to lose weight and turn off the ones that have been making you gain weight. Let's look at exactly how you can make this six-step process work for you.

Step 1: Eliminate the Causes of Thyroid Problems

The first thing to do is to carefully consider things that may interfere with your thyroid function and eliminate them. Certain foods have started to gain a reputation for playing a role in thyroid dysfunction, but this reputation isn't necessarily connected to the scientific evidence available.

Soy foods and the broccoli family (broccoli, cabbage, kale, Brussels sprouts, and collard greens) have all been said to cause thyroid dysfunction, but they also have many other health benefits. In addition, the research on these foods to date has been less than conclusive. In one study, rats fed high concentrations of soy had problems with their thyroid. The take-home message: if you are a rat, stay away from tofu. Human studies have shown no significant effect when soy is consumed in normal quantities.[2]

On the other hand, there are food groups for which there is substantive evidence supporting their link to an autoimmune disease of the thyroid that slows down your metabolism. Gluten is one of them.[3] If you think you are having a thyroid problem, you need to do a blood test (see chapter 11, Step 6) to identify any hidden reaction to gluten found in wheat, barley, rye, oats, kamut, and spelt.

Gluten sensitivity or allergy can cause many different types of symptoms, from migraines to fatigue to weight gain. Besides doing the blood test, you can simply eliminate gluten from your diet for three weeks. If your symptoms go away, you have a clue that your system might not like this food. If you want to take this self-test a step further, reintroduce gluten into your diet and see if your symptoms recur. If they do, that is another major clue.

This system is actually built into the UltraMetabolism Prescription. In the first three weeks of the program, I ask you to eliminate gluten from your diet. One of the reasons is that this will allow you to see for yourself if it is having a negative impact on your health and your ability to lose weight.

Testing yourself for food allergies is another important step to take and one that you will have to do in conjunction with a medical practitioner.

Toxins slow down your thyroid. Testing yourself for mercury and getting it out of your system and your environment are also essential.[4] Avoid fluoride, which has been linked to thyroid problems.[5] Don't drink chlorinated water. Checking for pesticides is more difficult, but supporting your body's detoxification system by eating organic foods, filtering your water, and eating detoxifying foods is helpful. See chapter 15 for more information on toxins and detoxification.

Stress also affects your thyroid function negatively. Military cadets in special forces training subjected to intense stress had higher levels of cortisol, higher inflammation levels, reduced testosterone, higher TSH, and very low T3. Treating the thyroid without dealing with chronic stress can precipitate more problems.

A common form of chronic stress known as adrenal gland exhaustion or burnout is particularly dangerous in this regard. Adrenal gland exhaustion occurs when your adrenal glands are unable to keep up with the physiological needs created by stress. Make sure to incorporate the relaxation exercises discussed in chapter 10 to reduce the effects of chronic stress on your thyroid and metabolism.

Step 2: Exercise and Take Saunas

There are two more elements that you need to incorporate into your lifestyle to treat your thyroid the right way: exercise and saunas.

Exercise stimulates thyroid gland secretion and increases tissue sensitivity to thyroid hormones throughout the body. The exercise regimen outlined in chapter 13 and again in chapter 16 is an excellent way not only to burn calories and boost your metabolism but to give your thyroid some additional support.

Besides being an excellent way to relax your muscles and your mind, saunas or steam baths are a good way to flush your system of pesticides that could be contributing to your thyroid problem. Saunas are an important aid to weight loss and thyroid repair because as you lose weight, fat tissue releases stored toxins such as PCBs and pesticides (organochlorines).[6] These toxins lower your T3 levels, consequently slowing your resting metabolic rate and inhibiting your fat-burning ability.[7] Therefore, detoxifying is an important part of improving your thyroid function. If you don't detoxify, your ability to lose weight *decreases* as you lose weight because of the negative effects of the released toxins on thyroid function.

Your whole body works as a system, a web, and attention needs to be paid to each of the 7 Keys to UltraMetabolism. One key might be more important than another for a particular person. We are not made of separate parts. That's why the doctor of the future will be a *supergeneralist,* instead of a super specialist. Once you get rid of the causes, improve your nutrition, exercise, reduce stress, and take the right supplements, if you still feel bad, can't lose weight, and have many low-thyroid symptoms, what can you do? The next step should be a trial of a prescription thyroid preparation. For this you will need to see a doctor experienced in treating subtle thyroid disorders.

> The doctor of the future will be a supergeneralist.

Identifying and correcting subtle thyroid problems is essential to a healthy metabolism and a healthy life. There are millions of undiagnosed and untreated people suffering from many common and vague complaints, including trouble losing weight despite eating a nutritionally rich diet and exercising. If you have any symptoms or suspect your thyroid may be a problem, find out more about testing and finding a doctor who can help at www.ultrametabolism.com.

Step 3: Eat Foods That Provide Nutritional Support for Your Thyroid, and Avoid Those That Don't

Every step on your road to healing and weight loss depends on proper nutrition and using food to communicate the right information to your genes.[8] Treating your thyroid is no exception to this. In addition to the guidelines offered in chapter 16, follow these suggestions to help heal your thyroid.

Eat Foods That Offer Nutritional Support for Your Thyroid

The production of thyroid hormones requires iodine and omega-3 fatty acids; converting the inactive T4 to the active T3 requires selenium; the binding of T3 to the receptor on the nucleus and switching it on require vitamins A and D and zinc. All of these elements are found in a good whole-food diet.

A clean, organic diet of whole foods will help provide the nutritional support you need. There are a few foods that are particularly good for thyroid problems:

- Seaweed and sea vegetables contain iodine.
- Fish (especially sardines and salmon) contains iodine, omega-3 fats, and vitamin D.
- Dandelion greens, mustard, and other dark leafy greens contain vitamin A.
- Smelt, herring, scallops, and Brazil nuts contain selenium.

Avoid Foods That Can Interfere with Thyroid Function

There are also a few foods that have been shown to inhibit thyroid function. To maximize the effects your diet has on your thyroid, you may want to avoid the following foods:

- **Gluten,** the protein found in wheat, barley, rye, oats, kamut, and spelt
- **Too much soy protein,** an excess (more than any average person would eat) of soy protein in the diet has been shown to

interfere with thyroid function only in patients who have
hypothyroidism and are being treated with thyroid medications.

Step 4: Use Supplements That Support Your Thyroid

The key nutrients needed for thyroid function are all included in the basic
supplement recommendations, including a multivitamin and mineral containing selenium, iodine, zinc, vitamins A and D, and the omega-3 fats (fish
oil), which are all critical to maintaining normal thyroid function.

One warning is that if your adrenal glands are burned out from long-
term stress, treating the thyroid without supporting the adrenal glands
through relaxation and adaptogenic herbs (such as ginseng, rhodiola, or
Siberian ginseng) can actually make you feel worse. Refer to the suggestions in the six steps in chapter 10 to help you balance your stress response.

Step 5: Have Your Thyroid Tested

There is no one perfect way, no one symptom or test result, that will properly diagnose low thyroid function or hypothyroidism. The key is to look at
the whole picture—your symptoms and your blood tests—and then decide. Doctors typically diagnose thyroid problems by testing your TSH levels and sometimes your free T4 level.

But some doctors and clinicians have brought the "normal" levels of
those tests into question. The diagnosis of "subclinical" hypothyroidism depends on having a TSH level over 5 mIU/ml and lower than 10. But new
guidelines from the American College of Endocrinologists suggest that
anything over 3 is abnormal.[9] This number is an improvement but may still
miss many people who have both normal tests and a malfunctioning thyroid system.

To get a complete picture, I recommend looking at a wider range of
function:

1. Thyroid-stimulating hormone (TSH) (the ideal is between 1 and
   2 mIU/ml)
2. Free T4 *and* free T3 (the inactive and the active hormone)
3. Thyroid antibodies (TPO), looking for an autoimmune reaction
   that commonly goes undiagnosed if the other tests are normal, as
   doctors don't routinely check this
4. Sometimes a test called the thyroid-releasing hormone (TRH)
   stimulation test

5. Or even the twenty-four-hour urine test for free T3, which can
   be helpful in hard-to-diagnose cases.

To a physician experienced in doing these tests, they can provide a fuller
picture of how your thyroid is functioning. The bottom line is that if you
think you have an undiagnosed thyroid problem, you should insist that your
doctor perform these tests or find a doctor who will. They are essential to fill
in the pieces of the puzzle that aren't given by the standard tests being used.

Step 6: Choose the Right Thyroid Hormone Replacement

Ultimately, to properly balance a thyroid that is severely out of balance, you
will need to go on some type of thyroid hormone replacement therapy.
There are certain things you can do by altering your diet and your lifestyle,
but if your thyroid isn't functioning properly you may need to take
some additional thyroid hormones to supplement its output. Knowing
what's available and what to ask about can empower you to make better decisions about your health.

When I went to medical school and did my residency, I learned about
only one treatment for low thyroid function: Synthroid, a synthetic form of
T4, a brand-name drug that wasn't even FDA-approved until recently. Why
do doctors prescribe it? Because that is all they are taught to prescribe. But
that doesn't make it the best treatment for everyone.

Many people benefit from this treatment. But in some cases the symptoms don't seem to go away using only T4, even if their tests return to normal. So what is the right treatment? The answer is . . . it depends. Part of the
beauty (and the headache) of the revolution happening in medicine now is
that there is no one right treatment for everyone.

A combination of experience, testing, and trial and error is necessary to
get any treatment just right. However, I have found that the majority of my
patients benefit from a combination
hormone treatment including both T4
and T3. Synthroid is just T4, the inactive
hormone. Most doctors assume that the
body will convert it to T3 and all will be
well. Unfortunately, pesticides, stress,
mercury, infections, allergies, and sele-
nium deficiencies can block that process. Since 100 percent of us have pesticides stored in our bodies, we will all likely have some problem with
Synthroid.

> The new medicine is personalized medicine—finding the right treatment for you.

The most common treatment I use is Armour thyroid,[10] a combination of thyroid hormones including T4, T3, and T2[11] (a little-known product of thyroid metabolism that actually may be very important). Armour is a prescription drug made from desiccated (dried) porcine thyroid. It contains the full spectrum of thyroid hormones, including T4, T3, and T2. It may seem paradoxical that taking a pig hormone can make you lose weight, but it does. The right dose ranges from 15 to 180 milligrams, depending on the person.

Many doctors still hold the outdated belief that the preparation is unstable and the dosage difficult to monitor. That was true with the old preparation of Armour, not the new one. (See www.armourthyroid.com for more information.) Sometimes the only way to find out if you have a thyroid problem is a short trial of something like Armour thyroid for three months. If you feel better, your symptoms disappear, and you lose weight, it's the right choice. Once started, it doesn't have to be taken for life (a common misperception). Sometimes, once all the factors that disturb your thyroid have been corrected, you may be able to reduce or discontinue the dose.

As with any treatment, always work with an experienced physician in using medications for your thyroid. Careful monitoring is essential. Taking

---

CHOOSING THE RIGHT MEDICATION

These medications must be used only under a physician's supervision. Options include:

Combination Preparations

- ❖ Armour thyroid
- ❖ Thyrolar
- ❖ Nature-Throid
- ❖ Westhroid

T4 Preparations

- ❖ Synthroid
- ❖ Levoxyl
- ❖ Levothroid
- ❖ Levothyroxine (generic)

T3 Preparations

- ❖ Cytomel
- ❖ Time-released T3

too much thyroid hormone or taking it if you don't need it can lead to un-desirable side effects, including anxiety, insomnia, palpitations, and, over the long term, bone loss.

Finding a Hidden Thyroid Problem

So many of my patients hope that the answer to their struggle with weight is because they have a "slow metabolism" or sluggish thyroid. While that is not true for everyone, it is true for about one in five women.

It is important to do some detective work and deal with factors that neg-atively affect the thyroid, particularly mercury and gluten; improve your nutrition; take a multivitamin and fish oil; and get the right tests and the right treatment if needed. This can truly be the answer to your health and weight problems.

I Have The Perfect Lifestyle—Why Can't I Lose Weight?

Amanda was a bright, active, successful, and determined young woman of 28. She was not overweight by conventional standards, with a BMI of 23 (remember, over-weight is >25 and obese >30) but wanted to lose the 20 extra pounds she carried well.

She worked hard at everything. Five days a week at 5:30 in the morning, she met with her trainer for forty minutes of aerobics and a half hour of weights. Her fitness level was excellent. She ate oatmeal, a protein shake, and a whole-wheat muffin or yogurt and fruit for breakfast. Lunch was salad, sliced turkey, and beans. She ate late but had only fish and vegetables. She didn't drink, smoke, or even eat sugar.

Doing all this, she understandably didn't feel satisfied with her difficulty in knocking off the extra weight. She should have been a twig. But when we tested her body fat, it was over 30 percent. How could this be?

Digging a little deeper into her story, I found some big clues overlooked by every doctor she saw. The biggest clue was that she never had a normal period and had been on birth control pills in order to menstruate regularly from a very young age. She also had very low blood pressure, was cold all the time, suffered from constipation, and was more tired that she should have been at 28.

We tested everything. All the tests were normal except for two: she had "low nor-mal" hormone levels for her thyroid tests and evidence of trouble burning fat in the mitochondria, discovered by a urine test of organic acids that identifies problems with metabolism (the thyroid hormone stimulates the mitochondria to burn more en-ergy and calories).

Other than suggesting that she eat dinner a little earlier, I advised no change in her diet or exercise program. All we added was the smallest dose of Armour thyroid, fish oil, a multivitamin and mineral to help her thyroid function, and carnitine and CoQ10 to help her mitochondria. I also told her to stop taking the birth control pills.

After three months she had a normal period for the first time in her life, felt more energetic, and had lost the extra 20 pounds. By treating the person and not the conventional thyroid lab tests, we were able to help her overcome a lifelong struggle with weight and give her back her fertility.

Summary

* Thyroid is the master metabolism hormone. If it is out of balance, your metabolism is out of balance.
* Hypothyroidism is one of the most underdiagnosed health issues in America. It is incredibly common, it has a huge impact on your health, and most doctors don't know how to treat it well.
* Diet, environmental toxins, and stress all affect thyroid function.
* There are a number of tests you can ask your doctor to do if you are concerned about hypothyroidism.
* The standard thyroid treatment, Synthroid, addresses only part of the picture. There are more sophisticated prescriptions available that will address your thyroid problems more completely.

CHAPTER 15

LOVE YOUR LIVER:

Cleansing Yourself of Toxic Weight

Toxins and Obesity: A Missing Link?

*J*oe worked for a major trucking company. In 1985, he contracted hepatitis C from a sexual encounter. In 1990, he first had problems with abnormal liver tests, but he didn't change his lifestyle much. He continued to smoke and drink.

His diet was typical of a hardworking middle-class man: eggs and sausage, doughnuts, a bagel with margarine for breakfast, meat sandwiches, cold cuts, and pepperoni pizza for lunch. Dinner was steak and wine, bread, potatoes, and pasta. Vegetables occasionally made an appearance.

He had stopped smoking and drinking a year before I first treated him, as his liver function and fatigue had worsened. He was under significant stress, with the recent loss of both his job and his wife. He was on injections of interferon, the latest medical treatment for his hepatitis, but his viral counts were still quite high. His blood pressure, blood sugar, and triglycerides were on the rise.

When he showed up in a workshop I teach called "Detox for Life" (see www. ultrametabolism.com/detox), he was ready for a change. And he was worried. His liver was scarring on a recent biopsy, things were starting to fail, and he was gaining weight around the middle. Not only did he have inflammation in his liver from the chronic hepatitis infection, but he also had a fatty liver from all the refined carbohydrates and junk food he consumed.

Joe was toxic. Years of smoking, drinking, and poor eating had poisoned his body, especially his liver. After five days on the detox program, which was a simple diet of vegetables, rice, beans, a detoxifying rice protein shake, a few herbs for the liver, and fiber to clean out the gut, along with saunas, yoga, and rest, he felt much better than he had in years.

After the program he came to see me, and I advised him to change his diet to control his insulin and blood sugar. I also recommended that he introduce some special foods such as broccoli, garlic, dandelion greens, and artichokes to boost his detoxification system and protect his liver. I added some special nutrients (N-acetylcysteine, alpha-lipoic acid, and selenium) and herbs (milk thistle) to further support his fatty liver.

After three months, he felt better, his weight had dropped by more than 20 pounds, and his viral counts were lower than they had been in years. Even with a compromised liver, healing can occur.

Environmental Toxins Cause Weight Gain

Toxins from within our bodies and toxins from our environment both contribute to obesity. Getting rid of toxins and boosting your natural detoxification system are essential components of long-term weight loss and a healthy metabolism. After years of practice and examining the scientific research, I have come to believe that these connections can no longer be ignored.

Could it be that environmental factors other than the changes in our diet and our sedentary lifestyle are contributing to the obesity epidemic?[1] The answer is, unfortunately, yes. The science of detoxification can help you overcome problems associated with exposure to the unprecedented levels of toxic environmental chemicals and heavy metals to which we are now exposed.

Let's see how toxic you are. Take the following assessment exam to figure out if this is a key you need to concentrate on.

HOW WELL IS YOUR DETOXIFICATION SYSTEM WORKING?

Score 1 point each time you answer "yes" to the following questions by putting a check mark in the box on the right. See page 78 for a reminder on interpretation of your score.

	YES
Are you constipated and go to the bathroom only every other day or less often?	☐
Do you urinate small amounts of dark, strong-smelling urine only a few times a day?	☐
Do you rarely break into a real sweat?	☐
Do you have one or more of the following symptoms: fatigue, muscle aches, headaches, concentration, or memory problems?	☐
Do you have fibromyalgia or chronic fatigue syndrome?	☐
Do you drink tap or well water?	☐
Do you have your clothes dry-cleaned?	☐

Do you work or live in a "tight" building with poor ventilation or windows that don't open? ☐

Do you live in a large urban or industrial area? ☐

Do you use household or lawn and garden chemicals or have your house or apartment treated for bugs by an exterminator? ☐

Do you have more than one or two mercury amalgams ("silver" fillings)? ☐

Do you eat large fish (swordfish, tuna, shark, tilefish) more than once a week? ☐

Are you bothered by one or more of the following: gasoline or diesel fumes, perfumes, new-car smells, fabric stores, dry cleaning, hair spray or other strong odors, soaps, detergents, tobacco smoke, or chlorinated water? ☐

Do you have a negative reaction when you consume foods containing garlic or onions, MSG, sulfites (found in wine, salad bars, dried fruit), sodium benzoate (a preservative), red wine, cheese, bananas, chocolate, or even a small amount of alcohol? ☐

When you drink coffee or other substances containing caffeine, do you feel wired, have increased aches in muscles and joints, or have hypoglycemic symptoms (anxiety, palpitations, sweating, and dizziness)? ☐

Do you regularly consume any of the following substances or medications: acetaminophen (Tylenol), acid-blocking drugs (Tagamet, Zantac, Pepcid, Prilosec, Prevacid), hormone-modulating medications in pills, patches or creams (the birth control pill, estrogen, progesterone, prostate medication), ibuprofen or naproxen, medications for colitis or Crohn's disease, medications for recurrent headaches, allergy symptoms, nausea, diarrhea, or indigestion? ☐

Have you had jaundice (turned yellow), or have you been told you have Gilbert's syndrome (an elevation of a liver test for bilirubin)? ☐

Do you have a history of any of the following conditions: breast cancer, smoking-induced lung cancer or other type of cancer, prostate problems, food allergies, sensitivities, or intolerances? ☐

Do you have a family history of Parkinson's disease, Alzheimer's disease, amyotrophic lateral sclerosis (ALS), or other motor neuron diseases or multiple sclerosis? ☐

If you don't want to write in this book, I've included this quiz in the companion guide that you can download by going to www.ultrametabolism.com/guide.

Answering yes to the questions posed above may alert you that you have an increased risk for exposure to toxins. But how common are toxins, and do they have anything to do with health, weight gain, or chronic disease?

Why should we worry about toxins? Unless we work with toxic chemicals or spray pesticides for a living, isn't our exposure minimal? I wish it were so, but it isn't. We live in a sea of toxins. Every single person and animal on the planet contains residues of toxic chemicals or metals in its tissues.

Since the turn of the twentieth century, 80,000 new chemicals have been introduced, and most have never been tested for safety. The Centers for Disease Control and Prevention (CDC) produced a report on human exposure to environmental chemicals that is very disturbing. They examined human blood or urine levels of 116 chemicals (and there were thousands more for which tests were not conducted) as part of the National Health and Nutrition Examination Survey (NHANES).[2]

They found toxic residues in almost every sample taken. Some had high levels of toxins; in many others the levels were low. Nonetheless, toxins were universally present. What's more, this study in isolation probably doesn't tell the whole story. Why? Because these chemical toxins move quickly from your blood into storage sites—mostly fat tissue, organs, and bones—so blood or urine levels severely *underestimate* our total toxic load.

Another disturbing set of data comes from the Environmental Protection Agency (EPA), which monitored human exposure to toxic environmental chemicals from 1970, when it began the National Human Adipose Tissue Survey (NHATS), to 1989. This study evaluated the levels of various toxins in fat tissue from cadavers and elective surgeries (now you know what happens to that fat they remove during liposuction).

Five of what are known to be the most toxic chemicals humans have created were found in 100 percent of all samples (OCDD, a dioxin, styrene, 1,4-dichlorobenzene, xylene, and ethylphenol—extremely toxic chemicals from industrial pollution that may cause harmful changes in your liver, heart, lungs, and nervous system). Nine more chemicals were found in 91 to 98 percent of the samples, including benzene, toluene, ethylbenzene, DDE (a breakdown product of DDT, the pesticide banned in the United States since 1972), three dioxins, and one furan. Polychlorinated biphenols (PCBs) were found in 83 percent of the population.

Another study based in Michigan found DDT in more than 70 percent of 4-year-olds. All of them were citizens of the United States, and, as mentioned above, DDT has been outlawed since 1972. How were these children exposed to these toxic chemicals? Probably through their mothers' breast milk. But then how did the moms acquire it?

With our global economy, we are often eating food that was picked a day or more before in Guatemala, Indonesia, or Asia, where there are not the same restrictions on the use of pesticides as there are in the United States. Many of these chemicals are stored in fat tissue, making animal products concentrated sources. One hundred percent of beef is contaminated with DDT, as are 93 percent of processed cheese, hot dogs, bologna, turkey, and ice cream, because the soil still contains residues of the pesticide long ago banned.

We are all stewing in this toxic soup, and there is little doubt that it is playing a major role in the current obesity epidemic in this country and people's ability to lose weight in general. But how? What are toxins? Where do they come from? How do they get inside us? Let's take a look.

What Are Toxins, and Where Do They Come From?

A toxin can be defined, in the broadest sense, as anything that doesn't agree with us. We can be exposed to toxic relationships, work environments, or our own toxic thoughts, which register in our body with a toxic stress response. But most of us think of toxins as poisons that contaminate our bodies in one way or another. These can be toxins that come from our normal bodily functions, and we know we eliminate these through our kidneys (via the urine) and our bile (via our liver and stool).

We generally cope well with these toxins except when our liver or kidneys fail. However, in the last hundred years we have been burdened with an unprecedented number of toxins, and the total load of all toxins—pesticides, industrial chemicals, mercury, and more—has exceeded our bodies' ability to get rid of them, leading to illness. It also, I believe, contributes to metabolic problems that promote weight gain and prevent weight loss.

Where do these toxins come from?

They come from two places. One is the environment (external toxins). The other is our own gut (internal toxins). The by-products of our metabolism (internal toxins) need to be processed. All of them put stress on our livers (more on this on the following pages).

External toxins include chemical toxins and heavy metals. The heavy metals that cause the most ill health are lead, mercury, cadmium, arsenic, nickel, and aluminum. The chemical culprits include toxic chemicals and volatile organic compounds (VOCs), solvents (cleaning materials, formaldehyde, toluene, benzene), drugs, alcohol, pesticides, herbicides, and food additives.

While most drugs are not truly toxins, certain medications can have toxic effects and cause weight gain, and can be considered external toxins.

Psychotropic medications, in particular MAO inhibitors, lithium, valproate, Remeron, Clozaril, Zyprexa, and in some cases, selective serotonin reuptake inhibitors (SSRIs), such as Prozac, Zoloft, and Paxil, have all been shown to promote weight gain through various mechanisms.

Billions of dollars are being poured into obesity drug research to find a magic pill that will burn fat or reduce appetite. It is clear that foreign chemicals such as medications can affect our weight and may have a role for some people. If drugs can affect our weight, then certainly other foreign chemicals, including environmental toxins, can cause weight gain.

Internal toxins include microbial compounds (from bacteria, yeast, or other organisms) and the by-products of normal protein metabolism (such as urea and ammonia). Bacteria and yeast in the gut also produce waste products, metabolic products, and cellular debris that can interfere with many body functions. These toxic elements can lead to increased inflammation and oxidative stress. These bacteria produce endotoxins, toxic amines, toxic derivatives of bile, and various carcinogenic substances, such as putrescine and cadaverine (you can imagine why they have those names).

All of these toxins affect your ability to lose weight. Because we store most of the toxins in our bodies, gaining weight itself has a kind of toxic effect. When you burn off the fat, the toxins come out, and if they aren't processed properly they can cause additional problems. In addition, our total toxic load can frustrate attempts at weight loss by impairing two key metabolic organs—the liver and the thyroid—and by damaging our energy-burning factories—the mitochondria.

Thoroughly scared now? I'm not giving you this information to frighten you but to make you aware of what is going on inside you and around you so that you can see how important it is to minimize your exposure to toxins and maximize your excretion of them. But before I get into teaching how to do that, let's look a little more carefully at exactly what toxins do to your body, particularly your liver, thyroid, and mitochondria.

Toxins: Disrupting Weight Control Signals and Turning Off the Metabolic Engine

What is it that prevents us from moving forward in our weight loss goals and interferes with our metabolism? A study I discussed briefly in chapter 14 is worth reviewing again. The title of the study is "Energy Balance and Pollution by Organochlorines and Polychlorinated Biphenyls,"[3] published in *Obesity Reviews* in 2003. The conclusion is that pesticides (organochlorines) and PCBs (from industrial pollution) are released from the fat tissue, where they are typically stored, and poison our metabolism, preventing us from losing weight.

The authors conclude that we should lose a *little* weight to reduce our risk of cardiovascular and degenerative diseases, but not too much because we could poison our metabolism. If there were no way to get rid of the toxins, I would agree, but there *are* many ways to "love your liver," or get rid of toxins.

How exactly do these chemical toxins interfere with metabolism? The researchers in the above-mentioned study, Pelletier, Imbeault, and Tremblay, reviewed sixty-three scientific studies on the link between chemical toxins and obesity and described many mechanisms.

First, people with a higher BMI (body mass index) store more toxins because they generally have more fat. Those toxins interfere with many aspects of metabolism, including reducing thyroid hormone levels and increasing excretion of thyroid hormones by the liver. Toxins also compete with thyroid hormones by blocking the thyroid receptors (the sites on the cell where thyroid hormones affect metabolism), and by vying for thyroid transport proteins, so the thyroid hormones have no way of getting to their site of action in the first place (something like stealing your car).

The bottom line: pesticides and other industrial toxins (PCBs) lower thyroid hormone levels, interfere with their function, and consequently slow metabolic functioning.

In a second study, a group of researchers from Laval University in Quebec found that those who released the most pesticides (organochlorines) from their fat stores during weight loss had the slowest metabolism after weight loss.[4] Their explanation for the decreased fat burning (also known as thermogenesis), after taking into account all other possible factors, was the exposure to pesticides.

In yet another study, the increase in toxins during weight loss in men inhibited normal mitochondrial function and reduced the subjects' ability to burn calories, retarding further weight loss.[5] This suggests that not only do toxins affect thyroid levels, they also damage mitochondria and reduce their ability to burn fat and calories. All of these actions cause both weight gain and resistance to weight loss.

Besides directly lowering thyroid hormone levels, breaking down your mitochondria, disturbing your metabolic rate, and inhibiting fat burning, toxins can damage the mechanisms by which hormonal signals control your appetite and eating behavior. Remember, *leptin* is the hormone that tells your brain you are full. Toxins (heavy metals such as mercury or chemical toxins) block these signals. Over time your brain becomes resistant to the effects of leptin, and you are hungry all the time. So exposure to toxins can increase your appetite.

All this is proof positive that toxins have a serious effect on whether or not you gain weight and your ability to lose it. But there's more. Toxins have also been proven to disrupt the highly orchestrated internal dance that is necessary to keep your body running smoothly and allow you to lose weight.

Hormone Disrupters: Hormonal Chaos

The dance of hormones, as we have seen, is critical to balancing your metabolism. Environmental chemicals and heavy metals are well-known hormonal disrupters. A Tufts University professor, Sheldon Krimsky, in his book *Hormonal Chaos: The Scientific and Social Origins of the Environmental Endocrine Hypothesis,* has extensively reviewed the research in this field.

He has found that low levels of these toxins—far below those considered acceptable by the Environmental Protection Agency—interfere with our normal hormone balance, including sex hormones, which may lead to early puberty in girls and an increase in hormonal disorders. Toxins can affect many of the major weight control hormones besides the thyroid. These include estrogens, testosterone, cortisol, insulin, growth hormone, and leptin.

In addition, toxins interfere with our stress response (our autonomic nervous system) and alter the normal circadian rhythms that control our eating behavior. These connections were explored at a conference sponsored by the National Institute of Environmental Health Sciences and Duke University entitled "Obesity: Developmental Origins and Environmental Influences."[6]

While there is still much to learn about the connection between weight gain and toxins, we can no longer ignore their impact. It is certainly not the only factor in our obesity epidemic—or in any one person's struggle with weight—but it must be considered part of the equation. That is why learning how to minimize your exposure to toxins and to use food, exercise, supplements, and even saunas is so important in helping you attain Ultra-Metabolism.

Fatty Liver: Cause or Effect of Weight Gain?

There is one more major detrimental health problem that too many toxins in your body cause: fatty liver. It is a big problem, because the more problems you have with your liver, the harder it is for you to process toxins of any kind. Let's look at what a fatty liver is and what it means for your health and your ability to lose weight before we start exploring ways to get these terrible toxins out of your body.

Have you ever tasted foie gras, the wonderfully tender, juicy, and fatty French treat served in the world's best restaurants? *Foie gras* is French for "fatty liver" (everything sounds better in French). Force-feeding ducks or geese starchy carbohydrates (corn) induces their livers to become laden with fat. Unfortunately, it's not just the ducks that suffer from fatty liver. Fatty liver is the most common liver disease in America, affecting 20 percent of the population.

Is the cause drugs, a virus, or pollution? None of those. It's caused by the most abundant toxin in our diet: *sugar.* Increases in sugar or refined carbohydrate consumption, as we have seen, increase insulin and insulin resistance. This leads to the accumulation of fat in your liver cells. Increased fat inside the liver cells is produced by eating too much sugar, refined carbohydrates, and high-fructose corn syrup, which causes insulin resistance. Some of my patients have developed cirrhosis of the liver from sugar alone!

The sugar is turned into triglycerides (the fat in fatty liver), which then fills up the liver cells (and other cells). Excess sugar calories also increase oxidative stress and further damage the energy factories, or mitochondria. Poisoned mitochondria can't effectively burn fat or calories, which leads to a slower metabolism and more weight gain—the vicious cycle experienced by millions of people.

If that weren't bad enough, having a fatty liver impairs detoxification further. A fatty liver is also *inflamed,* which leads to nonalcoholic steatohepatitis (NASH), a form of hepatitis caused by insulin resistance. A fatty liver produces more inflammatory molecules and free radicals and leads to even more mitochondrial damage. Full of fat and inflammation, the liver can no longer protect you from the damaging effects of other environmental toxins, which leads to even further damage.

Your Ability to Detoxify: How Your Genes Control Your Detoxification System

The effect of toxins on any individual is determined, in part, by the genetics of that person's detoxification systems. Some people can mobilize and eliminate toxins well; others have more difficulty. Getting rid of heavy metals is important. It depends on specific proteins and enzymes that bind to the metals and transport them out of the cells.

In one recent study, mice bred without the protein (metallothionein) that is necessary for heavy metal detoxification gained more weight over their lifetime than mice that could eliminate the metals.[7] This suggests that there are genetic factors that predispose you to being able to rid toxins from your body easily. But even if you are genetically predisposed to problems with detoxification, there are things you can do to cleanse your body of these terrible materials and limit their effects on you.

Priming Your Detoxification System: Keeping Yourself Clean and Thin

Reviewing how toxins block your metabolism, interfere with your weight control mechanisms, disrupt your hormones, damage the mitochondria, increase inflammation and oxidative stress, lower your thyroid hormones, negatively affect your circadian rhythms and the autonomic nervous system, and generally lock up all the keys to UltraMetabolism has likely made you feel depressed. For that I apologize. Unfortunately, I didn't have any input into the design of human beings, and I do try my best not to contribute to environmental pollution (I even traded in my SUV for a compact car).

But there is good news: by making some simple lifestyle choices, you can reduce your exposure to toxins and dramatically increase your ability to mobilize and get rid of stored toxins. First, I will provide a little background on how your detoxification system functions. In the book I coauthored, *UltraPrevention* (Simon and Schuster, 2003; find more at www. ultraprevention.com), and in the *The Detox Box* (Sounds True, 2004; find more at www.drhyman.com/DetoxBox.aspx), I review this system in more detail.

Even if you are genetically predisposed to having problems with detoxification, you can improve your detoxification system by following the recommendations in this chapter. If you scored in the moderate to high range on the quiz at the beginning of this chapter, follow this six-step program to help you improve the effectiveness of your detoxification system and your weight loss efforts.

- **Step 1:** Minimize your exposure to toxins.
- **Step 2:** Sweat it out and sweat it off.
- **Step 3:** Eat foods that help you detoxify, and avoid foods that make you toxic.
- **Step 4:** Try herbal remedies that will help you detoxify.
- **Step 5:** Add supplements to support your liver and help you detoxify.

✤ **Step 6:** Consider testing your detoxification system and looking for toxins.

If you are having problems with this UltraMetabolism key and you have too many toxins in your body, you can reverse the effects of your toxic environment by following these steps. By doing this you can customize the UltraMetabolism Prescription for your own particular needs. This will allow you to turn on genes that cause you to lose weight and turn off the ones that have been making you gain weight. Let's look at exactly how you can make this six-step process work for you.

Step 1: Minimize Your Exposure to Toxins

Avoiding toxins when you can is important. While toxins are nearly ubiquitous in our culture, there are a number of easy, practical things you can do to avoid them. You don't need to be hypervigilant to diminish the amount of toxins you are exposed to. Try some of the following strategies. You don't need to do everything at once. Try to add one suggestion at a time. The two most important are eating organic foods when possible and drinking clean, filtered water.

- ✤ **Eat organic foods.** Eat food and animal products raised without the use of petrochemical pesticides, herbicides, hormones, and antibiotics (see Resources).
- ✤ **Drink filtered water.** Reverse-osmosis and carbon filters are the best systems for detoxifying your water.
- ✤ **Use air filters.** HEPA/ULPA filters and ionizers can be helpful in reducing dust, molds, volatile organic compounds, and other sources of indoor air pollution.
- ✤ **Clean and monitor your heating systems.** Eliminate the release of carbon monoxide, the most common cause of death by poisoning in America.
- ✤ **Keep houseplants.** They help filter the air.
- ✤ **Air out your dry cleaning.** Do so after you bring it home, before wearing it.
- ✤ **Avoid excess exposure to environmental petrochemicals.** These include garden chemicals, dry cleaning fumes, car exhaust, and secondhand smoke.
- ✤ **Reduce or eliminate the use of toxic household and personal care products.** Underarm deodorant, antacids, and pots and pans containing aluminum are good examples.

- ⁜ **Remove allergens and dust.** Cleanse your home as much as possible of these elements.
- ⁜ **Minimize electromagnetic radiation** (EMR). EMR from radios, TVs, and microwave ovens can be toxic.
- ⁜ **Reduce ionizing radiation.** Don't overdo sun exposure or medical tests such as X-rays.
- ⁜ **Reduce your exposure to heavy metals.** These can be found in predatory fish (tuna, swordfish, shark, tilefish, sea bass), river fish, water, lead paint, and thimerosol-containing products such as vaccines.
- ⁜ **Avoid cooking with Teflon-coated pans.**
- ⁜ **Have one to two bowel movements a day.**
- ⁜ **Drink six to eight glasses of water a day.**
- ⁜ **Exercise regularly.** Yoga or lymphatic massage can improve lymph flow and help flush toxins out of your tissues into your circulation so they can be detoxified.

Step 2: Sweat It Out and Sweat It Off

Regular use of a sauna or steam bath may impart a similar stress on the cardiovascular system [as exercise], and its regular use may be as effective as a means of cardiovascular conditioning and burning calories as regular exercise.

—W. DEAN, "Effect of Sweating,"
Journal of the American Medical Association
1981, vol. 246, p. 623.

Eating all the right foods, taking the right supplements (see pages 202–203), and minimizing your exposure to toxins are all good ideas, but how can you prevent stored fat toxins from harming you as you lose weight?

The answer: Take a sauna or steam bath. Heat therapy is a significantly underutilized treatment in medicine. It helps balance the autonomic nervous system, reduces stress, lowers your blood sugar level, helps you burn calories, and helps to excrete pesticides and heavy metals through your skin.

Sometimes used as a treatment for chemical poisoning, saunas were used to help detoxify the workers at Ground Zero in Manhattan after 9/11. While more research is needed, a review paper on thermal therapy suggests many promising effects, including reduction of inflammation and oxidative stress,[8] as well as weight loss.[9]

In a two-week study of twenty-five obese adults, body weight and body fat were reduced after sauna therapy for fifteen minutes at 60 degrees Celsius daily in a far-infrared sauna. Researchers reported on another obese patient who couldn't exercise because of knee arthritis and lost 17.5 kilograms, decreasing body fat from 46 percent to 35 percent after ten weeks of sauna therapy.

Weight loss is inhibited when you have high levels of toxins. This is, in part, because the toxins block thyroid function. When you get rid of these toxins, you keep your metabolism going faster. Saunas or steam baths assist in this process. They help you release the toxins through your skin as you sweat. In this way you stop storing toxins that would cause your metabolism to slow down.

Sauna therapy or heat therapy has many benefits. I recommend regular use of this ancient and effective tool for maintaining your health.

Step 3: Eat Foods That Help You Detoxify, and Avoid Foods That Make You Toxic

Our detoxification system relies on the right balance of protein, fats, fiber, vitamins, minerals, and phytonutrients to be effective. This is built into the general food guidelines for the UltraMetabolism Prescription. Eating the right foods can prime your system for detoxification. Eating the wrong foods can block your body's ability to detoxify. There are a few special foods worth mentioning here that can be particularly effective in boosting your detoxification system.

Eat Foods That Help You Detoxify

Food plays a major role in priming your system to excrete toxins. For example, adequate protein is required to supply the amino acids used by the liver to provide the building blocks for the powerhouse of our detoxification system, *glutathione*. Glutathione is the most critical antioxidant and detoxifier made by the body—and one that is easily depleted in the face of chronic exposure to toxins. If you don't eat enough protein, you won't be able to produce enough of this important detoxifier.

Phytonutrients are another example. These special molecules found in plant foods boost your detoxification pathways. These include many colorful plants and foods included in the UltraMetabolism Prescription.

The following foods will help you prime your system for detoxification:

- Cruciferous vegetables (broccoli, kale, collards, Brussels sprouts, cauliflower, bok choy, Chinese cabbage, Chinese broccoli)
- Green tea
- Watercress
- Dandelion greens
- Cilantro
- Artichokes
- Garlic
- Citrus peels
- Pomegranate
- Cocoa

Step 4: Try Herbal Remedies That Will Help You Detoxify

Some herbal remedies help your body in its detoxification process. The best herbal supplement for your liver is milk thistle. And add green tea to your daily regimen.

Try these herbs to improve your detoxification:

- Milk thistle (silymarin)
- Green tea (drink it or take it as a supplement)

Step 5: Add Supplements to Support Your Liver and Help You Detoxify

You might want to try using the following supplements to assist you in your detoxification process.

- **Extra-buffered vitamin C with mineral ascorbates.** During any state of toxicity, vitamin C needs are increased.
- **N-acetylcysteine.** This important amino acid derivative helps to boost glutathione in and protect the liver from everyday toxic stresses.
- **Amino acids such as taurine and glycine.** Amino acids are some of the key molecules that the liver uses to help it detoxify.
- **Alpha-lipoic acid**
- **Bioflavonoids (citrus, pine bark, grape seed, green tea).** These compounds are the key plant compounds or pigments, about four thousand in total, that provide the color to plants. They include:

- **Quercitin**
- **Pycnogenol or grape seed extract**
- **Rutin**
- **Probiotics** such as *Lactobacillus* and *Bifidobacter* species help normalize gut flora and reduce endotoxins (toxins produced by imbalances in the gut bacteria).

Step 6: Consider Testing Your Detoxification System and Looking for Toxins

Sometimes doing a little detective work to find out where your genetic detoxification weak spots are and learning how to support them with diet and supplements can be very helpful. Finding hidden toxins is also important. Mercury is probably the most important toxin to look for if you are chronically ill. Here are a few of the tests that may be worth exploring with an experienced physician.

- Genetic testing of detoxification pathways (SNPs): special testing for the genes that regulate our ability to eliminate toxins
- Measurement of detoxification enzymes
  - Glutathione peroxidase
- Heavy metals
  - RBC or whole blood levels of metals; this usually indicates only a recent exposure.
  - Hair analysis, which identifies mercury mostly from fish consumption, not dental fillings.
  - Chelation challenge with DMPS or DMSA are medications that help bind to the metals in your body and then help you excrete them in the urine, where they can be measured

Detoxify for a Healthy Metabolism

Our detoxification system depends on specific foods, phytonutrients, exercise, adequate sweating, vitamins, minerals, accessory nutrients (or "conditionally essential" nutrients occasionally needed in excess of the body's capacity to synthesize them), hyperthermic or heat treatments such as saunas or steam baths, and stress management to function at optimal levels.

Occasionally, specific medical treatment is necessary to test for and treat toxins related to problems such as mercury or lead toxicity. You don't have to do every detox method at once, but doing simple things can make a big difference over the long term: eat organic food (especially animal products);

include foods such as broccoli, green tea, watercress, and artichoke in your diet; choose fish with lower mercury content (such as wild salmon and sardines—see www.epa.gov/waterscience/fishadvice/advice.html, the U.S. Environmental Protection Agency Fish Advisory; filter your water; exercise regularly; take a good multivitamin and mineral; and last, sweat it out!

Blissful Mexican Weight Loss

One cold winter, I took a group of twelve people to Mexico for a week of detoxification. But it was anything but deprivation.

The resort, Maroma, was on the Mayan Riviera. The sand was soft and white, the sun was warm, and the thatched roofs and white stucco buildings were serene. At night, white candles lit the paths.

We all came from different places, from different worlds, looking for a respite, a way to stop life for a brief week and return to health. Each person came with different health concerns and conditions, but for one week we shared a collective experience designed to find the "pause" button for our lives and our nervous systems.

We all ate a cleansing, detoxifying diet, Mexican style—fresh vegetables, beans, whole-grain rice, fish, simple rice protein shakes, and soothing detoxifying broths. The meals were delicious, incorporating healing local plants such as chia (once eaten by the Pima Indians).

We had something to eat five times a day and shared our meals with laughter and pleasure. In the mornings we woke early, sat quietly watching the sunrise, then walked and talked on the beach. We did yoga, had classes, and rested our nervous systems.

In the evenings we sat in the temazcal, a Mayan steam bath. We all unplugged from the world for a week, from all our compulsive habits and addictions—whether to Starbucks or cyberspace. The transformations in each person, in just one week, were remarkable.

Weight loss was a by-product of switching to whole, healing foods and was done without hunger or a sense of deprivation; some people lost a few pounds, others 8 to 12 pounds. But more important, everyone detoxified from stress, from doing. A few cell phones rang, but when we left, all our faces were less swollen, tense, and fatigued.

Nourishing the body and soul is simple. A few basic principles can provide healing: whole clean food, things to soothe the nervous system like yoga, following a regular rhythm of waking, sleeping, eating, and exercise, steam baths and nature, and letting go of toxic habits in community with laughter and tears.

Summary

- Toxins are a universal problem in the world we live in. They are something you need to concern yourself with both for your health and for weight loss.
- Toxins inhibit the function of your thyroid and your mitochondria as well as throwing your hormones out of balance, all of which wreak havoc on your metabolism.
- Some people are genetically predisposed to detoxify more easily than others. The good news is that even if you have genetic problems with detoxification, you can do some simple things that will make detoxification easy.

PART III

The UltraMetabolism Prescription

⁘

In this final part of the book I am going to teach you the two phases of the UltraMetabolism Prescription. Think of this not so much as a diet but as a way of life. Incorporating the principles of the UltraMetabolism Prescription into your everyday life will help you develop the UltraMetabolism you have been looking for and lose weight quickly and easily. The two phases of the program are:

1. Detox your system
2. Rebalance your metabolism and maintain a lifelong healthy metabolism

In addition to the delicious recipes and shopping lists you will find in this part of the book, I am going to help you integrate all the information you gathered from the 7 Keys so that you can personalize and optimize this program based on your personal needs. This will allow you to have the owner's manual for your body you have been looking for. You will learn exactly how to use the herbs and supplements I recommended you try in the various keys, you will learn how to exercise smarter, and you will learn how to change your diet so that you are eating healthy whole foods that are delicious, give you energy, keep you young, and send messages to your genes that help you lose weight.

But remember, this isn't a mechanistic regimen, this is your life. These aren't rules that you have to follow without fail day in, day out. These are guidelines for eating and living that are designed to work with your genes rather than against them. To stay on the prescription, you will need to be flexible and open. Find out what works for you, and use that information to get fit and stay healthy for life.

CHAPTER 16

THE ULTRAMETABOLISM PRESCRIPTION OVERVIEW:

A Science-Based Whole-Food Approach to Eating

The UltraMetabolism Prescription: A Two-Phase Program

The UltraMetabolism Prescription is a two-phase program preceded by a one-week preparation phase. The first phase lasts three weeks and will help you resolve metabolic problems; it involves a period of cleansing and renewal through detoxification. You will feel more energetic, lose weight, relieve many chronic health problems, and improve your energy, memory, digestion, sleep, and allergies.

The second phase lasts for a lifetime. It will help you live in harmony with your genes and rebalance your hormones, immune system, and energy metabolism and maintain lifelong healthy weight and UltraMetabolism.

Integrating the Menus with the 7 Keys

By combining the menus and recommendations for choosing healthy foods, along with the suggestions for exercise, supplements, relaxation, and saunas and steam baths at the end of each UltraMetabolism key in part II, not only will you lose weight and learn simple, effective ways to change your habits and reset your metabolism, but you will build the foundation for lifelong health.

Food as Medicine

In the Chinese language the word for eating is made up of the characters for "eat rice," or *chi fan*. The characters for "take medicine" are *chi yao*. In Chinese culture, eating food and eating medicine are synonymous. Food is medicine.

Modern science also teaches us to use food as medicine. The delicious, simple, nourishing menus in the UltraMetabolism Prescription use real, traditional, whole foods and were created on principles derived from the current collective scientific wisdom as well as nutritional knowledge of our evolutionary diets. This is the diet on which our bodies were designed to thrive.

The translation of this science of weight loss and UltraMetabolism into practical, delicious, and nourishing meals and recipes comes from the collective culinary creativity of Kathie Swift, M.S., R.D., the former nutrition director of Canyon Ranch Lenox for more than a decade, and me. We have decades of experience working with stressed and time-pressured people struggling to lose weight.

The recipes and suggestions in the prescription have been designed to fit into a busy life. They generally have short preparation and cooking times so that you can prepare a delicious, healthy meal on a limited time schedule. But keep in mind that even the fastest meals cannot be made without some forethought and organization. Most of us don't plan our meals. Then we suddenly end up in an alarm state, hungry and on the hunt. We have become a nation of drive-by eaters. This is the worst way to eat if you want to get healthy and lose weight.

The steps of the program are laid out in day-by-day, easy-to-follow menus and suggestions that will help you find the key to unlock the door to UltraMetabolism and weight loss without struggle, deprivation, or suffering. Travel and snack menus are integrated into the program so you are never stuck without the right balance of foods to control your metabolism. Shopping lists and resources are provided, so you can find everything you need to succeed. All of these resources and the shopping lists are also available on www.ultrametabolism.com/guide.

All of this is offered to make the program as easy to follow as possible. Nonetheless, you are going to have to put some effort into it. Turning your metabolism into UltraMetabolism will require planning, shopping, and preparation. The energy you put into creating these wonderful dishes will pay off.

In addition, you will need to be open-minded. You are going to discover new ingredients that offer potent nutritional benefits and wonderful flavors and help create a healthy metabolism. Some of the ingredients won't be familiar to you. Remember, we live in a culture that has convinced us that unhealthy foods are good for you.

Some of the familiar has to be abandoned to rebuild a diet that is healthy and will help you lose weight. Be open to trying new things. I think you will find them quite delicious. And I am sure you will find they give you more energy and make you feel better than many of the foods you were eating before.

Learning to incorporate these new foods into your diet and figuring out where to find them and how to prepare them takes time. Be patient. Food has become the enemy for many. Don't let food be your enemy. It is time now to make friends, to celebrate food, and to discover something new. Eating well is not impossible despite the cultural and marketing pressures that surround us. Food is an adventure down the road to your birthright: feeling fabulous and fit.

When you follow the program, food will once again become a friend, a source of nourishment, a pleasure to eat, and a delight to share with friends and family. We all have a fit person inside of us waiting to emerge. We all can fit into our jeans if we learn how to fit into our genes by feeding them what they deserve. The UltraMetabolism Prescription will teach you how to do that.

An Overview of the UltraMetabolism Prescription

To get you ready for the journey ahead, I would like to review the phases of the program in a little more detail.

Preparation Phase: Let Go of Bad Habits (One Week)

We are often locked into habits without recognizing their effect on us. Most of us are drug addicts and don't realize it. Sugar, junk food, caffeine, and alcohol all affect our ability to function, and though they temporarily make us feel better, they often deplete us in a deeper way.

Taking a "drug holiday" is an important way to discover how you really feel and give yourself the opportunity to tune in to your own body's signals for hunger, sleep, and relaxation. Getting off sugar, HFCS, hydrogenated fats, junk food, alcohol, and caffeine for this one-week preparation phase, in fact, is powerful enough to totally change your health and help you lose weight quickly. Take a risk. If you do nothing else, doing this will change your life.

Phase I: Detoxify Your System (Three Weeks)

In this phase you are going to clean up your diet. Getting rid of garbage foods, moving toward a diet of whole, unprocessed foods, and eliminating foods that you may have sensitivities to will help you start the weight loss process and reboot your metabolism. This phase sets the stage for the rest of the diet. In it people typically lose 6 to 11 pounds and start to feel energetic and healthier. This detoxification phase is designed to further what you started in the preparation phase by removing the most common food aller-gens—gluten, dairy products, and eggs—and introducing whole, healing foods.

Beside weight loss, you can expect to feel more energetic, to sleep better, to be rid of chronic sinus and digestive problems and headaches. Part of the healing occurs because of all the junk and common food allergens you are eliminating, but most occurs because you introduce delicious, whole foods.

Phase II: Rebalance Your Metabolism and Maintain a Lifelong Healthy Metabolism (Four Weeks to Life!)

During Phase II of the program you will start to reeducate your body and program your genes to lose weight and keep it off. By doing so you will lose an additional 5 to 10 pounds in the first two to four weeks and then approx-imately a pound a week until you are at your optimal weight. You will stick to the whole-food diet eating plan you started in Phase I. But you will rein-troduce all of the foods that have the potential for intolerances.

When reintroducing them, you can monitor their effects on your health. (If you get a stomachache or your nose stuffs up when you eat dairy prod-ucts, it is best to stay away from them.) This phase solidifies the hormone and immune changes in your body and allows you to reset your metabolism for the long term.

Phase II of the program is really just a start for the rest of your life. Variety, fun, nourishment, pleasure, color, and wholeness are essential for making this way of eating your way of eating for life. Feel free to improvise and adapt; just stick with whole foods, and you will have difficulty get-ting into trouble. By the end of eight weeks (one week preparation, three weeks of the detoxification phase, and four weeks committed to re-balancing your metabolism for life) you will have lost 11 to 21 pounds or more and feel so much better that you won't even want to eat all the foods that you used to love but that ruined your metabolism. UltraMetabolism is now yours.

Customizing the UltraMetabolism Prescription

In part II you were given a series of step-by-step recommendations to help you customize the UltraMetabolism Prescription to fit your body's own unique needs. As you start on the prescription, bring these techniques with you if you wish to. The prescription by itself is a powerful way to get health-ier and lose weight. But if you want to turn up the power of the program, use what you learned in part II to customize the prescription.

You don't have to use every technique in every chapter. First look back at the quizzes and limit your customization to places where you are currently having difficulty. If this still looks as though it is too much to do, simply take a few of the steps or techniques in those chapters and try using them. You don't have to rush into everything at once. Over time you can gradually include more and more of these techniques in your daily routine. You should do what works for you to make you feel healthier and help you lose weight.

If you are only going to take a few of the steps in the chapter, try starting with the first few steps for each customization program you want to use. In most of the keys, one of the steps revolves around food, and those principles are included in the UltraMetabolism Prescription. Steps 4, 5, and 6 are about herbs, supplements, and testing, and you need to include these items only if you wish to. (There is more on pages 217–218 on how to use herbs and supplements.) The herbs and supplements in each chapter have been shown to help people optimize their metabolism and overcome problems in certain areas, and I strongly encourage people to take them if they have identified problem areas. But this is not a part of the program that you absolutely *have* to do.

If you concentrate on only one thing in this book, concentrate on the food program in this part of the book. Use the keys in part II as necessary to customize and personalize the program. The keys will help you unlock the door to what may have been lifelong struggles with health and weight.

Basic Food Guidelines for UltraMetabolism

Important Message for Those Who Own a Body

The menus and recipes are provided for your pleasure and convenience, but the real benefit of UltraMetabolism is to learn the principles of eating that will nourish your soul, satisfy your stomach, control your appetite, and help you permanently lose weight and achieve optimal health. I have provided for you here the lessons learned over a lifetime of work with patients, and derived from the latest scientific research on how to make your genes say *"Yes!"* Study these principles and experiment. Other than these guidelines, there are no fixed rules about carbohydrates, protein, fat, ratios, or calories. Listen to your body. It will thank you!

The UltraMetabolism Prescription: General Principles

Resources for finding special foods, snacks, shakes, or hard-to-find organic animal products can be found at www.ultrametabolism.com/guide.

Meal Timing

- **Include protein at breakfast every day,** such as whole omega-3 eggs, a protein shake (soy) such as UltraMeal Plus, nuts, seeds, or nut butter.
- **Eat something every three to four hours** to keep your insulin and glucose levels normal.
- **Eat small snacks** including protein, such as a handful of almonds or other nuts or seeds and a piece of fruit in the morning and afternoon.
- **Avoid eating for two to three hours before going to bed.** If you have a snack earlier in the day, you won't be as hungry even if you eat a little later.

Meal Composition

- **Control the glycemic load of your meals.** This is very important. You can do this by combining adequate protein, fats, and whole-food carbohydrates from vegetables, legumes, nuts, seeds, whole grains, and fruit at every meal or snack. It is most important to avoid eating quickly absorbed carbohydrates alone, as they raise your sugar and insulin levels.

Travel Suggestions

- **Almonds** in a Ziploc bag are a useful emergency snack. One handful is a good portion for a snack. This can be eaten with a piece of **fruit.** Real food is the best.
- **Food bars (there are many, but here are some of my favorites)**
    - **Omega Smart:** Apricot and Almond, Chocolate Nut Bars, Cinnamon Apple, or Carrot Cake (high in fiber, protein, and omega-3 fats, gluten- and dairy-free, and made from real food)
    - **Unibar:** Chocolate Cherry or Blueberry Almond (higher in protein, gluten-, and dairy-free, made from whole foods) made by North American Pharmacal
    - **Biogenesis:** UltraLean Gluco Balance Bars in Chocolate, Spice, Crispy Rice, and Berry (gluten- and dairy-free and high in protein and nutrients)

See www.ultrametabolism.com for food bar sources.

Basic UltraMetabolism Whole Real Food Principles

Choose from a variety of the following foods:

- **Choose organic produce and animal products** whenever possible.
- Cold-water **fish** such as salmon, halibut, and sable contain an abundance of **beneficial essential fatty acids,** omega-3 oils that reduce inflammation. A great source of smaller, wild Alaskan salmon, sable, and halibut high in omega-3 fats and low in toxins is Vital Choice seafood (www.vitalchoice.com). Canned wild salmon is a great emergency food.
- **Eat high-quality protein such as fish,** especially fatty, cold-water fish such as salmon, sable, small halibut, herring, and sardines, plus shellfish.
- **Eat omega-3 eggs,** up to eight a week.
- **Create meals high in low-glycemic legumes** such as lentils, chickpeas, and soybeans (try edamame, Japanese soybeans in a pod, quickly steamed with a little salt, as a snack). These foods slow the release of sugars into the bloodstream, helping to prevent excess insulin release, which leads to hyperinsulinemia and its related health concerns, including poor heart health, obesity, high blood pressure, high LDL ("bad") cholesterol, and low HDL ("good") cholesterol.
- **Eat a cornucopia of fresh fruits and vegetables** teeming with phytonutrients—carotenoids, flavonoids, and polyphenols—which are associated with a lower incidence of nearly all health problems, including obesity and aging.
- **Use more slow-burning low-glycemic vegetables** such as asparagus, broccoli, kale, spinach, cabbage, and Brussels sprouts.
- Berries, cherries, peaches, plums, rhubarb, pears, and apples are **optimal fruits.** Organic frozen berries (Cascadian Farms) can be used in your protein shakes.
- **Focus on anti-inflammatory foods** including wild fish and other sources of omega-3 fats, red and purple berries (these are rich in polyphenols), dark green leafy vegetables, orange sweet potatoes, and nuts.
- **Eat more antioxidant-rich foods,** including orange and yellow vegetables, dark green leafy vegetables (kale, collards, spinach,

etc.), anthocyanidins (berries, beets, pomegranates), and purple grapes containing trans-resveratrol, blueberries, bilberries, cranberries, and cherries. In fact, antioxidants are found in all colorful fruits and vegetables.

- **Include detoxifying foods in your diet,** such as cruciferous vegetables (broccoli, kale, collards, Brussel sprouts, cauliflower, bok choy, Chinese cabbage, and Chinese broccoli), green tea, watercress, dandelion greens, cilantro, artichokes, garlic, citrus peels, pomegranate, and even cocoa.
- **Use herbs** such as rosemary, ginger, and turmeric in cooking that are powerful antioxidants, anti-inflammatories, and detoxifiers.
- **Avoid excessive quantities of meat.** Use lean organic or grass-fed (when possible) animal products in moderation—beef, chicken, pork, lamb, buffalo, ostrich, and so on. Good sources are whole-food or other local health food stores (also see mail order sources).
- **Garlic and onions** are noted for their cholesterol blood-pressure-lowering and antioxidant effects. They are also anti-inflammatory and enhance detoxification.
- **A diet high in fiber** further helps to stabilize blood sugar by slowing the absorption of carbohydrates and supports a healthy lower bowel and digestive tract. Try to gradually increase fiber to 30 to 50 grams a day and include soluble or viscous fiber (legumes, nuts, seeds, whole grains, vegetables, and fruit), which slows sugar absorption from the gut.
- **Use extra virgin olive oil,** which contains anti-inflammatories and antioxidants. It should be your main oil.
- **Organic soy products** such as soy milk, soybeans, edamame, and tofu are rich in antioxidants that can reduce cancer risk, lower cholesterol, and improve insulin and blood sugar metabolism.
- **Increase your intake of nuts and seeds,** including raw walnuts, almonds, macadamia nuts, and pumpkin and flaxseeds.
- And yes . . . **chocolate,** only the darkest, most luxurious kind and only one to two ounces a day. It should be at least 70 percent cocoa.

Decrease your intake of (or ideally eliminate):

- **All processed or junk foods.**
- Foods containing **refined white or wheat flour and sugar,** such as breads, cereals (corn flakes, Frosted Flakes, Puffed Wheat,

and sweetened granola, etc.), flour-based pastas, bagels, and pastries.

- All foods containing **high-fructose corn syrup.**
- All **artificial sweeteners** (aspartame, saccharin, etc.).
- **Starchy, high-glycemic cooked vegetables,** such as mashed white potatoes.
- **Processed fruit juices,** which are often loaded with sugar. (Instead, try juicing your own carrots, celery, and beets, or other fruit and vegetable combinations.)
- **Processed canned vegetables** (which are usually very high in sodium).
- Foods containing **hydrogenated or partially hydrogenated oils** (which become trans fatty acids in the bloodstream), such as most crackers, chips, cakes, candies, cookies, margarine, doughnuts, peanut butter, processed cheese, and so on.
- **Processed, refined oils** such as corn, safflower, sunflower, cotton-seed, peanut, and canola.
- **Red meats (unless organic or grass-fed) and organ meats.**
- **Large predatory fish and river fish,** which contain mercury and other contaminants in unacceptable amounts, including swordfish, tuna, tilefish, and shark.
- **Dairy products.** Substitute unsweetened, gluten-free soy milk, almond milk, or hazelnut milk products.
- **Caffeine.** Limit as much as possible (try to switch to green tea or have a half cup of coffee a day).
- **Alcohol.** Limit to no more than three glasses of red wine per week.

Exercise

As I mentioned in chapter 13, exercise is the only thing besides eating breakfast that has been correlated with long-term weight loss. We were built to move, and I feel exercise is important and healthy. I do not encourage you to sit on the sofa all day, even if you are doing the UltraMetabolism Prescription.

That being said, exercise is an optional part of this program. The exercise program laid out in chapter 13 is a powerful way to stay fit and help you lose weight even if you didn't score high on the quiz in that chapter. The concepts about exercise found in chapter 13 are useful to everybody, and if you are looking for an exercise regimen to follow, you may want to consider what is said there.

But there are ways that you can exercise without really "exercising." You don't have to go to the gym, run on a treadmill, and pump iron to stay in shape. Just start moving around more. Go for walks with your friends or family. Go out and do some gardening. Play Frisbee in the park with your kids. Pick up a tennis racket and just knock a tennis ball around. You don't even have to play a real match if you don't want to. Anything that you can do to get out and move your body can be considered exercise. So don't think that you absolutely have to go to the gym to get fit. Just use your body more.

A fun way to check how much you are exercising is to get a step counter. These devices are relatively inexpensive, and they are a great way to get a sense for how much you move around every day. If you buy one, see if you can get up to 10,000 steps per day. Whatever you do, just get moving!

Herbs and Supplements

In addition to the basic nutritional supplementation I recommend on page 218 that is appropriate and recommended for *everyone,* in each chapter in part II I have given you a list of optional herbal remedies and supplements that you can take to improve your metabolism and help you overcome certain conditions. Details about these are found in Appendix B.

Again, these are not a critical part of the program. But I have treated many patients who have benefited from the various herbal remedies and supplements discussed in those chapters. You can put that information to use if you wish to.

When you start to take herbal remedies and supplements, there are a few things that you need to consider.

Not All Herbs and Supplements Are Created Equal

Be aware that all brands are not created equally. Quality is up to the manufacturer because of limited regulations regarding manufacturing. Certain companies are more careful about quality, sourcing of raw materials, consistency of dose from batch to batch, the use of active forms of nutrients, not using fillers, additives, colorings, etc.

When choosing supplements, it is important to choose quality products. See www.ultrametabolism.com/guide for more information on choosing quality supplements.

Finding the best products to support health has always been the most difficult part of my job. The lack of adequate government regulations, the dizzying number of products on the market, and the large variations in quality all create a minefield of obstacles for anyone trying to find the right vitamin or herb.

Fortunately, in a sea of poor quality and lowered standards, there are a few companies that stand out and have stepped up to the responsibility of producing safe and effective products. They meet my specific criteria for quality and effectiveness.

While I have tried to make educated judgments about companies and their products, I am unable to verify all claims about every product. Therefore each person must be cautious and evaluate companies and products for him- or herself. I offer this only as part of my hard-won knowledge about how to evaluate supplements.

1. Look for GMP (good manufacturing practices) drug or supplements standards from an outside certifying body.
2. Try to verify third-party analysis for active ingredients and contaminants (see www.consumerlab.com).
3. Try to use products that have some basis in basic science or clinical trials or have a long history of use.
4. Use clean products, free of fillers, binders, excipients, flow agents, shellacs, coloring agents, gluten, and lactose.

While I do not officially endorse or have any consulting or employee relationship with any supplement companies, I do believe a few have risen to the top of the supplement industry and can be safely used to help support and enhance your health.

Unfortunately, many of these products are designed for therapeutic use by physicians, nutritionists, and other health care practitioners and are unavailable to the average consumer. Please see www.ultrametabolism.com/guide for more information on finding quality supplements.

The Lenox Village Integrative Pharmacy has researched many of the best-quality supplements on the market. You can receive guidance from expert pharmacists and order by phone at (888) 796-1222.

Basic Nutritional Support

Ninety-two percent of Americans are deficient in one or more essential vitamins and minerals, and more than 99 percent of Americans are deficient in the essential omega-3 fatty acids. Therefore I recommend that *all* people take a basic multivitamin and mineral, calcium, magnesium with vitamin D, and omega-3 fats as the foundation for good health, as well as a healthy metabolism. Ample scientific evidence supports this recommendation, including guidelines published in the *New England Journal of Medicine* and the *Journal of the American Medical Association*.

Before starting the supplements listed in part II that relate to specific conditions, I recommend that everyone follow the recommendations for basic nutritional support. Nearly all people need three basic supplements to create and maintain good health and an optimal metabolism. The food we eat today simply does not give us all of what we need in terms of vitamins and minerals. So many parts of your metabolism depend on vitamins, minerals, and essential fats that you cannot have an optimal metabolism without them.

If you are interested in supplements and want to get started on a supplement regimen now, consider the following:

1. A multivitamin and mineral combination
2. A balanced, absorbable calcium, magnesium, and vitamin D supplement
3. An omega-3 fatty acid supplement

A Multivitamin and Mineral Combination

A good multivitamin and mineral generally contains the following:

- ❖ Mixed carotenes (alpha, beta, cryptoxanthin, zeaxanthin, and lutein), 15,000–25,000 units
- ❖ Vitamin A (preformed retinol), 1,000–2,000 international units
- ❖ Vitamin D3, 400–800 international units
- ❖ Vitamin E (mixed tocopherols, including D-alpha, -gamma, and -delta), 400 international units
- ❖ Vitamin C (as mixed mineral ascorbates), 500–1,000 milligrams
- ❖ Vitamin K1, 30 micrograms
- ❖ Vitamin B1 (thiamine), 25–50 milligrams
- ❖ Vitamin B2 (riboflavin), 25–50 milligrams
- ❖ Vitamin B3 (niacin), 50–100 milligrams
- ❖ Vitamin B6 (pyridoxine), 25–50 milligrams
- ❖ Folic acid (ideally as mixed with folic acid 5-methyltetrahydrofolate), 800 micrograms
- ❖ Vitamin B12, 100–500 micrograms
- ❖ Biotin, 150–1,000 micrograms
- ❖ Pantothenic acid, 100–500 milligrams
- ❖ Iodine, 25–75 micrograms
- ❖ Zinc (as amino acid chelate), 10–30 milligrams

- Selenium, 100–200 micrograms
- Copper, 1 milligram
- Manganese, 5 milligrams
- Chromium (ideally as chromium polynicotinate), 100–200 micrograms
- Molybdenum, 25–75 micrograms
- Potassium, 50–100 milligrams
- Boron, 1 milligram
- Vanadium, 50 micrograms
- Inositol, 25–50 milligrams
- Choline, 100–200 milligrams

Keep in mind that this usually requires the intake of two to six capsules or tablets a day to obtain adequate amounts. Some people may have unique needs for much higher doses that need to be prescribed by a trained nutritional or functional medicine physician. (See www.ultrametabolism.com/guide for definition and for finding a functional medicine physician.)

A Balanced, Absorbable Calcium, Magnesium, and Vitamin D Supplement

In addition to a multivitamin and mineral, you will need to consider taking additional calcium, magnesium, and vitamin D supplements. They are usually found packaged together in one supplement. I recommend the following:

- Calcium citrate, 800–1,200 milligrams per day
- Magnesium amino acid chelate (aspartate, glycinate, ascorbate, or citrate), 400–600 milligrams per day
- Vitamin D3, 400–800 international units per day (in addition to what is in the multivitamin because so many people are significantly vitamin D–deficient)

An Omega-3 Fatty Acid Supplement

Finally, I recommend supplementing your intake of omega-3 fatty acids. They are so difficult to come by in our modern diet that supplements help almost everyone. Try the following:

- EPA/DHA (approximately 400 mg/200 mg ratio per capsule), one to four capsules a day (this must be from a reputable company that certifies purity from heavy metals and pesticides; safer options include Nordic Naturals, OmegaBrite, Metagenics)

Additional Supplements

Once you have started on this basic nutritional supplement program, you can begin to integrate the various other herbal remedies and supplements that are discussed in each of the keys. When you do this, keep a few things in mind:

1. **Don't double-dose.** Many supplements help in a number of different keys. Don't double-dose. Simply take the highest dosage recommended for a given supplement. Doing this will help in each of the areas connected to that supplement. For example, if I suggest 600 milligrams a day of lipoic acid to help balance your blood sugar, 600 milligrams a day to improve your mitochondrial function, and 600 milligrams a day to help your detoxification, it does *not* mean you should take 1,800 milligrams a day, only a total of 600 milligrams.
2. When taking herbs, be careful to select companies that provide purified and well-processed herbs. Not all herbs are created equally, and some contain a significant amount of contaminants.
3. Take fish oil just before meals to prevent any fish taste from coming up.
4. In general, take all your vitamins with food—optimally with a meal or just before. People who take them after a meal may find they just sit on top of their food and upset their stomach. If you still have an upset stomach, find a doctor who can help correct any digestive problems, which are often the source of intolerance.

Introduction to Menus and Recipes: Before You Start Cooking

By now you should understand that UltraMetabolism is about empowering you in the care and feeding of your body. It is about nourishment, relaxation, and pleasure, about understanding how your system functions and learning the tools to help heal and repair a dysfunctional metabolism. (If you have skipped forward to this section without reading the previous chapters, you will miss all the secrets of how your body works and have to rely on the recipes instead of your understanding.) These menus and recipes incorporate the nutritional wisdom and intelligence that will provide the foundation for a lifetime of improvisation. Food will become your ally, not your enemy.

UltraMetabolism is not a diet (although it can be used to lose weight). It is a way of living and eating that nourishes and heals the body at the deepest level.

Exploring the aisles of a new food store, browsing the abundant sources of good food at the local farmer's market or your neighborhood Asian grocer, or on the Internet with the many organic food sites provided in the Resource list, you will create a nourishing food experience to integrate into your UltraMetabolism lifestyle. You will be surprised to find foods you never imagined could be so enjoyable.

Endless Culinary Choices

The UltraMetabolism Prescription will lead you on the path to endless culinary choices, many of which will supply you with the basic raw materials for a healthy metabolism and life. If you prefer certain ingredients, use those. If you particularly like one recipe and don't like another, then substitute. If you want to mix and match days within each phase, go ahead. Feel free to eat any Phase I recipe in Phase II. These are guidelines, not rules.

Food should provide a sense of abundance and nourishment, as well as wonderful sensory experience, not one of deprivation. Experimenting, learning, and being creative are the keys to self-care and optimal health. Discover what is good for *you*.

Summary

- ❖ Food is medicine. The UltraMetabolism Prescription teaches you how to take the most up-to-date scientific information and use food as medicine for yourself.
- ❖ The UltraMetabolism Prescription is an easy-to-follow eight-week, two-phase program that will help you detoxify your body, rebalance your metabolism, and maintain an optimum metabolism for life.
- ❖ You can customize the UltraMetabolism Prescription by using the information that you learned in part II. Doing this will elevate the program a level and help you turn your metabolism into UltraMetabolism.
- ❖ You don't have to exercise to exercise. Instead of going to the gym, try going for a walk, gardening, playing with your kids, or going for a hike.

❖ Take the basic nutritional support—a multivitamin and mineral supplement; calcium and magnesium with vitamin D; and omega-3 fatty acids (fish oil).

❖ Herbal remedies and supplements can be a powerful way to help you customize the program. Use the information in this chapter and in part II to help you if you have specific problems with one of the keys to UltraMetabolism.

❖ The UltraMetabolism Prescription isn't a diet; it is a way of eating for life. Use it to become healthy, lose weight, and cook foods that are enjoyable to you.

CHAPTER 17

HOW TO CREATE YOUR OWN ULTRAKITCHEN:

Getting Started: The Empowered Kitchen

If you were climbing a mountain or planning a trek into unknown territory you would make sure you had the right clothing and tools. You would study the map before you set out. The journey of self-discovery, the one that will take you to a healthy metabolism, also requires some preparation and equipment. Some special tools and instructions will make your journey successful.

Let's start with your kitchen. Having the basic equipment makes food preparation easier and faster, as will learning what to remove from your pantry and what hazards to avoid in the marketplace. The suggestions and guidelines offered here will provide practical tools for your adventure into a healthy metabolism—an UltraMetabolism.

Arm Yourself with the Proper Tools

Consider this equipment a tool kit for taking care of your body. You can substitute or make due with other tools if need be, but I would strongly recommend that you consider purchasing the following items if you don't already have them.

I would also recommend that you buy the best-quality tools as you build your kitchen. After all, if you were climbing a mountain you would buy boots that would last for the duration. The items in this list are as vital to your health as an excellent pair of boots would be if you were to go mountain climbing. These tools can last you a lifetime if you start with ones that are quality items and take proper care of them.

I consider the following to be the basic essential hardware for the care and feeding of a human being (or at least the feeding)!

- A set of good-quality knives
- Wooden cutting boards—one for animal products, another for fruits and vegetables
- An eight-inch nonstick sauté pan

- A twelve-inch nonstick sauté pan (nonstick pans can vary in quality; buy the highest quality, such as Cephalon or All Clad, because of the health risks of poorer-quality nonstick pans using Teflon)
- An eight-quart stock pot
- A two-quart saucepan with lid
- A four-quart saucepan with lid
- An eleven-inch-square nonstick (non-Teflon) stovetop griddle
- Three to four cookie or baking sheets
- A food processor
- A blender
- An immersion blender
- A chef's thermometer
- A can opener
- A coffee grinder (for flaxseeds)
- Wire whisks
- Spring tongs
- A fish spatula
- Rubber spatulas
- Parchment paper (natural)
- Assorted measuring cups: one quart, one pint, and one cup, both dry and liquid style
- A lemon/citrus reamer
- Microplanes (assorted sizes)

Rid Your Kitchen of Fat-Burning Enemies

Before you purchase the items you need to start the program, take an afternoon to cleanse your cabinets of the items that are harmful to your health and metabolism. This includes eliminating toxic fats and sugars from your cabinets so that you won't "accidentally" add them to a recipe.

Start by throwing into the trash all items containing hydrogenated and partially hydrogenated fats and high-fructose corn syrup. Those two changes alone can radically alter your life by changing your cells and your metabolism. Reading the label (see chapter 9) on each of the food products in your cabinets and refrigerator will tell you which ones have these products in them.

Creating Your Very Own Ultrakitchen: Tips and Tricks

Over the years, working with the best nutritionist in the country, Kathie Swift, I have accumulated some tips and tricks to eat well and feel great. I offer them to you here.

Read the Shopping List and Learn the Best Brands

By eliminating what you don't need anymore, you will open up space in your refrigerator and cabinets for more healthy alternatives. In the section of recipes that follows, I will provide you with weekly shopping lists so you can easily go to the market and restock your cabinets with what you need. Read the shopping lists and brand recommendations for products and familiarize yourself with higher-quality foods that do not contain food additives.

Local food co-ops or national chains such as Whole Foods, Trader Joe's, and Wild Oats have nearly all the foods and brands recommended in the UltraMetabolism Prescription. Be proactive and urge your supermarket chain to carry these types of products. Buy foods that have not been or are only minimally processed. I recommend choosing organic foods whenever possible to reduce your exposure to pesticides, and increase your intake of vitamins, minerals, antioxidants, and phytonutrients.

Choose Organic, Hormone- and Antibiotic-Free Food

Buy antibiotic- and hormone-free animal products—dairy products, poultry, and red meat—whenever possible. Avoid eating fish that contain high levels of mercury, such as swordfish, tilefish, shark, king mackerel, and fresh tuna (canned tuna, especially chunk light, is lower in mercury). I recommend you eat fish with the least mercury, including blue crab (mid-Atlantic), flounder, sole, salmon (wild), sardines, herring, anchovies, and shrimp. Check the periodic updates on seafood safety on www.ewg.org and www.oceansalive.org.

Buy a variety of seasonally fresh, locally grown, and, whenever possible, certified organic produce. Though organic food is generally more expensive, the benefits are worth it. Organic foods do not contain the high levels of pesticides, hormones, and antibiotics found in conventional foods. Research indicates that organic foods also have more nutrients than foods grown conventionally. Some non-organic produce has much higher levels of pesticides.

The following is a priority list for purchasing organic produce based on data from the Environmental Working Group (www.ewg.org):

- Strawberries
- Cantaloupe
- Grapes

- Cherries
- Apricots
- Peaches
- Apples
- Pears
- Cucumbers
- Celery
- Red pepper and green peppers
- Spinach

Some non–organically grown items in your local grocery store are still relatively healthy to eat. If you can't completely stick to organically grown produce (either because your grocery store doesn't carry it or because the cost is prohibitive), the following eleven items are generally considered the products that are least contaminated by pesticides. While I encourage you to buy as much organic produce as you can, if you can't these are items you would probably be safe purchasing in a conventionally grown form.

- Asparagus
- Avocados
- Bananas
- Broccoli
- Cauliflower
- Kiwis
- Mangos
- Onions
- Papayas
- Pineapples
- Peas (sweet)

Check out the Environmental Working Group Web site, www.ewg.org, for further updates. You can also reduce your exposure to pesticides and bacteria by washing your produce well. Prepare a vegetable wash solution using 1 teaspoon mild soap or 1 tablespoon cider vinegar in one gallon of water. Wash your vegetables in this solution and rinse well. Use a vegetable brush on potatoes, sweet potatoes, carrots, and other hard produce whose skin you plan to eat.

Seek Out These Antioxidant Powerhouses

Scientists continue to learn more about measuring antioxidants in foods, referred to as oxygen radical absorbance capacity (ORAC), or the ability of the food to soak up damaging free radical molecules in a test tube. Be aware that with ongoing research ORAC values may change and new foods may be added or change places on the list.

In the meantime, have fun with this Top 20 list of antioxidant foods. Be sure to include plenty of these on your shopping lists. How many of these foods do you like? Were there any that surprised you by making the Top 20, such as russet potatoes, maligned by many popular diet books? Which ones might you now be more likely to introduce?

1. Small red beans, dried
2. Wild blueberries
3. Red kidney beans
4. Pinto beans
5. Blueberries, cultivated
6. Cranberries
7. Artichokes, cooked
8. Blackberries
9. Prunes
10. Raspberries
11. Strawberries
12. Red Delicious apples
13. Granny Smith apples
14. Pecans
15. Sweet cherries
16. Black plums
17. Russet potatoes, cooked
18. Black beans
19. Plums
20. Gala apples

DRINK CLEAN WATER

Contact your local water department to find out about the quality of your drinking water. Consider a water-purifying system such as a reverse osmosis filter for your home. (See www.ultrametabolism.com for approved water filters.) If you drink bottled water, choose glass or clear, hard, durable plastic containers (versus soft, opaque, thin, easily bendable plastic). Soft plastics tend to release toxic chemicals, including phythalates and bisphenol A, which have been linked to hormonal disorders and infertility.

Forgo These Foods Forever

Avoid food products that contain:

- Hydrogenated or partially hydrogenated oils
- High-fructose corn syrup
- Artificial sweeteners such as aspartame, saccharin, acesulfame-K, cyclamate, neotame, saccharin, sucralose, etc.
- Sugar alcohols such as sorbitol, mannitol, xylitol, and maltitol, which often cause gas and intestinal upset
- Artificial fats such as olestra
- Artificial colorings (dyes, such as FD&C Yellow No. 5, No. 3, etc.)
- Preservatives such as BHA and BHT
- Brominated vegetable oil (BVO), a known toxic additive that is found in some citrus sodas
- Heptyl paraben (a preservative used in beer and some noncarbonated drinks)
- Hydrogenated starch hydrolysate (a sweetener)
- Hydrolyzed vegetable protein (a flavor enhancer added to instant soups, sauce mixes, and hot dogs)
- Monosodium glutamate (a flavor enhancer added to many foods but shown to cause reactions in some people, perhaps by over-stimulating brain activity)
- Propyl gallate (a preservative found in edible fats, such as mayonnaise, oils, shortening, baked goods, and dried meats)
- Potassium bromate (a flavor enhancer found in breads and banned in several countries as a carcinogen)
- Sodium nitrite and sodium nitrate (preservatives found in processed meats that have been linked to cancer)
- Sulfites (sulfur dioxide and sodium bisulfite, preservatives found in wine, dried fruit, instant potatoes, French fries, pizza, and other foods, linked to headaches and severe allergic reactions in some)

Check out the Center for Science in the Public Interest's Web site, www.cspinet.org, for further information and updates on "chemical cuisine."

Eat These Foods Sparingly

You can eat these foods from time to time (once or twice a month after the first four weeks of the UltraMetabolism Prescription), but make sure you do not eat them too often.

- White flour and white flour products. "Wheat flour" or "enriched wheat flour" is essentially the same as white flour, unless the label explicitly lists "whole wheat flour" as the first ingredient.
- Refined sugars. Refined sugar is most often sugar separated from the stalk of sugar cane or from the beet root of the sugar beet. The sugar-containing juice is extracted, processed, and dried into sugar crystals.
- Highly saturated animal fat in meats and dairy products (fatty meats, deli meats, sausage, etc., as well as dairy fat).
- Alcohol (should be avoided altogether by sensitive individuals).

Suggestions for Success

While you prepare and cook your meals, keep these suggestions in mind. They can make cooking a relaxing and enjoyable experience, allowing food to become your ally instead of your enemy and your kitchen to become a sanctuary instead of a battleground. With a little patience and practice, you will feel very comfortable in the kitchen.

- Get organized, think through the week ahead, and take one day each week to spend a few hours shopping and cooking.
- Review your shopping list carefully before you head to the store.
- When you come home from the store, organize your groceries in the refrigerator and pantry.
- Don't be intimidated by the recipes; simply read over each one carefully before starting to cook.
- Put on some fun music and wash and cut up vegetables on the weekend or the day before and store in Ziploc bags in the refrigerator. Have them all ready to go. You are much more likely to eat them if they are all ready to cook.
- Multitask while in the kitchen: simmer some soup or cook a grain for the next day while preparing dinner.
- Also, double or even triple the recipes and freeze some for later use. Having a meal ready to go in the freezer is as good as having money in the bank!

Building the Perfect Kitchen

Remember that your kitchen is one of the most important rooms in your home. It is the place where you prepare the food that is going to nourish and sustain you and your family. When your kitchen is out of balance or you don't have the right tools, it is difficult to prepare healthy meals that turn on the genes that keep you healthy and make you lose weight.

As you start on the UltraMetabolism Prescription, try to implement some (if not all) of the tips in this chapter so that your kitchen and the time that you spend preparing food can be enjoyable and rewarding. Getting re-enchanted with nourishing, delicious food, with eating and pleasure is so important—it is at the center of human life and belongs in an ultrakitchen, not the front seat of your minivan!

Summary

❖ Your kitchen utensils are your tool kit for your journey on the UltraMetabolism Prescription. Buy good tools to make the journey more rewarding.

❖ Cleanse your kitchen of foods that are harming you and replace them with quality organic whole foods that will nourish you and help you lose weight.

❖ When choosing packaged food, make sure you read the label carefully. Checking to see what is in the products you are purchasing is an important step in cleansing your kitchen and your body.

❖ Make cooking relaxing and fun by reviewing your menus and recipes before you start cooking, planning ahead, and listening to music you like while you are in your kitchen.

CHAPTER 18

HOW TO AVOID PITFALLS AND COMMON CHALLENGES

Taking UltraMetabolism to Work and on the Road

You shouldn't let frequent travel, being out of the house, or working at mealtime get in your way of health and UltraMetabolism. When you travel, it can be somewhat difficult to find quality food options that will fit into the UltraMetabolism Prescription. When you go to work, it can be daunting to have to prepare and pack a lunch every day.

What follows is a number of menus that you can easily take with you. You can either prepare these on the road or prepare them in advance and bring them with you in a cooler. They will help you stay out of trouble in the vast minefield of convenience stores and junk food that awaits you when you leave home.

Grab-and-Go Options and Prep and Pack

These grab-and-go options give you some quick choices when you are really busy and don't have time to prepare a whole meal and still allow you to eat healthy, whole foods. You can use all these starting with Phase II. These foods can be prepared at home in advance. Consider keeping a small cooler in your car so you are never without healthy choices. Keeping in mind our busy lives, I often suggest canned beans or legumes, or frozen vegetables. Feel free to cook dried beans or chop fresh vegetables yourself.

BREAKFAST CHOICES

Grab-and-Go Breakfast 1: Yogurt Parfait
- Organic plain, nonfat, or low-fat yogurt or whole soy yogurt
- Fresh or frozen berries
- Milled flaxseed (Bob's Red Mill or Arrowhead Mills)
- Chopped raw nuts such as almonds, walnuts, or pecans

Grab-and-Go Breakfast 2: Morning Wrap

- ⁘ Tofu, baked (WhiteWave)
- ⁘ Sprouted grain tortilla (Food for Life or French Meadow Bakery or Food for Life brown rice wrap is a good gluten-free option)
- ⁘ Dijon mustard with curry powder
- ⁘ Fresh fruit in season

Grab-and-Go Breakfast 3: Eggs Dijon

- ⁘ Omega-3 eggs, hard-boiled the night before
- ⁘ Whole-grain rye bread slices
- ⁘ Yogurt Dijon (two parts plain organic soy yogurt mixed with one part Dijon mustard)
- ⁘ Fresh fruit in season

Grab-and-Go Breakfast 4: Nutty Banana

- ⁘ Banana
- ⁘ Natural nut or seed butter (almond, macadamia, cashew, sunflower)
- ⁘ Raw wheat germ (Bob's Red Mill or Arrowhead Mills)
- ⁘ Chopped nuts such as Brazil nuts or walnuts

Grab-and-Go Breakfast 5: Omega Morning

- ⁘ Sardines or wild salmon (Vital Choice)
- ⁘ Red onion, tomato, fresh dill sprigs
- ⁘ Whole-grain rye bread or crackers (Trader Joe's, RyVita, or Mestemacher & Seitenbacher)
- ⁘ Pink grapefruit

LUNCH CHOICES

Grab-and-Go Lunch 1: Soup 'n' Such

- ⁘ Lentil or bean soup (Walnut Acres, Westbrae, Shari's Organics, etc.)
- ⁘ Baby spinach salad with tahini dressing (or Drew's, Annie's, or Newman's dressing)
- ⁘ Flax crackers (Matter of Flax or Mary's Gone Crackers)
- ⁘ Small apple

Grab-and-Go Lunch 2: White Bean Wrap

- ⁘ Sprouted grain tortilla (Food for Life or French Meadow Bakery)
- ⁘ White cannellini beans (Westbrae, Eden)

- Fresh arugula leaves or fresh basil leaves
- Avocado and tomato
- Drizzle of extra virgin olive oil
- Fresh pear

Grab-and-Go Lunch 3: Wild Fish Roll
- Wild salmon or sardine salad: wild salmon or sardines, drained (Vital Choice)
- Fresh dill sprigs and red onion and watercress
- Soy mayonnaise (Spectrum) with a dab of horseradish if desired
- Sprouted grain roll (Alvarado St. Bakery or French Meadow Bakery)
- Blood orange

Grab-and-Go Lunch 4: Mediterranean Salad
- Mesclun greens (Earthbound Farms)
- Fresh mozzarella or Veganrella Mozzarella (nondairy option)
- Roasted red peppers
- Artichoke hearts
- Kalamata olives
- Walnuts
- Fresh lemon juice and extra virgin olive oil
- 15–20 red grapes

Grab-and-Go Lunch 5: Asian Soup and Salad
- Miso soup, 1 cup, with baked tofu (WhiteWave)
- Cabbage salad: ready and washed coleslaw mix of cabbage and carrots
- Thirty-second dressing: rice vinegar, toasted sesame oil, kelp powder, and sesame seeds or Drew's or Annie's Sesame Dressing
- Brown rice and seaweed crackers (Edwards and Son)

Eating Out Wisely

Eating out often leads to eating too much, and too much of the wrong things. But as awareness grows and the needs of health-conscious diners are met, menu options are changing and nutritionally intelligent choices are now available. Even some chain restaurants now offer healthy options.

A few simple suggestions about dining out can go a long way toward keeping you healthy and boosting your metabolism. Remember, there are no bad foods if they are whole foods. As Paracelsus, the ancient physician, told us, "The dose makes the poison."

KATHIE SWIFT'S BLESSINGS ON THE MEAL: THE 3/3/3 TECHNIQUE

To enjoy your meal and help yourself relax while you eat, try the 3/3/3 technique. The results of this simple technique can be remarkable. Awareness, breath, and nourishment are all critical to optimal digestion and metabolism.

1. Enjoy a moment of gratitude before you begin eating.
2. Take three relaxing deep breaths to the count of three in and three out before you begin eating.
3. During the first few minutes (about three minutes) of your meal, rest your hands in your lap at least three times, taking a relaxing breath with each break.

- **Prepare your mind** before you begin your meal by practicing the 3/3/3 Technique. This is especially important when dining out because you may not always be able to control the menu's quality, but by eating with awareness, you will be more likely to control your overall intake.
- **Discover some "slow food" restaurants** where the atmosphere and ambience are soothing to your senses. Our eating environment influences how much we end up eating.
- **Enjoy ethnic cuisine,** including Indian, Japanese, Thai, Mediterranean (Italian, Greek, and Spanish), and Middle Eastern. You will reap different nutritional benefits from traditional ingredients such as lemongrass in Thai dishes, sea vegetables in Japanese dishes, curry in Indian dishes, and great greens such as escarole and broccoli rabe in Mediterranean dishes.
- **Be inquisitive.** Ask questions about ingredients, and don't be afraid to ask for substitutions in a dish. For example, instead of white rice, request brown rice; instead of a starchy fried vegetable, double the green vegetable.
- **Mix and match menu items.** If you see a dish on the menu that comes with cranberry orange relish and your dish comes with gravy, ask the waiter to switch. If your dish has a vegetable you don't particularly like while another dish has a vegetable you adore, "mix and match" to get the healing foods you love.
- **Request a "crudités platter," fresh fruit, or olives** as a starter or appetizer instead of the bread basket. Bread and alcohol at the beginning of a meal increase your hunger and alcohol decreases

your inhibitions, making it more likely you'll make a play for the
cheesecake.

- ⁘ **Request "double the veggies."** Restaurants may overload the
  quantity of fish, meat, and poultry and skimp on the produce, especially the steamed vegetables.
- ⁘ **Order a light drink.** Try a Virgin Mary, sparkling water with a
  spritz of lemon or lime, or an herbal tea instead of alcohol before
  the meal.
- ⁘ **Check in with your gut-brain before ordering dessert.**
  Rate your sense of satisfaction and if you are a "3-Gently satisfied" (1-Not satisfied, 2/3-Gently satisfied, 4/5-Too full), use your
  freedom of choice to skip dessert. There will always be another
  night for the perfect indulgence.

Eating with Your Family

The recipes and menus in the UltraMetabolism Prescription can be enjoyed by the whole family. Parents and children alike can benefit from a
whole, unprocessed, real-food diet. If you take the time to prepare nutritious meals for your family that you can sit down and enjoy together, you
will find that it benefits your life in many ways. One hundred years ago
nearly all meals were eaten at home; now half are eaten outside the home.

Very few families enjoy regular mealtimes together. Very few families
make the time to prepare and enjoy food and one another at the end of the
day. This is an ancient tradition that existed for a reason. We slow down, we
spend time with those we love sharing food and nourishment with. All of
this helps our metabolism. It might even help us all get along better.

Teaching your children to enjoy meals and conditioning their taste buds
to the delicious tastes of whole organic vegetables and grass-fed or organic
animals offers them the opportunity to understand food in a way that you
might not have had growing up in this country. Teaching your children
about the value of food is a way to give them a gift that will stay with them
their entire life. What's more, having your child help with food preparation
is an important experience and a lifelong skill for healthy living.

You are a busy person. We all are. But try not to fall prey to the designs of
corporate America's fast-food culture. If you share the menus and recipes in
the UltraMetabolism Prescription with your family and teach your children about the value of eating well, you may provide them with the opportunity to avoid the weight and health consequences you are suffering right
now. At the very least, if you sit and eat with your family at night, you will
give your family an understanding that food is important, that it is some-

thing to be shared and enjoyed with the people in your life, and that it is a vital way to maintain optimum weight and health.

Remember, food is information talking to our genes. Share that information with your family.

Summary

- It may be more difficult but is not impossible to eat the Ultra-Metabolism way even when you are at work or traveling. Some simple menus that you can take with you will keep you on the Prescription even when you are surrounded by junk food.
- Eating out is another potential pitfall. But with consumer demand changing, more and more restaurants are offering their customers healthy options. Keeping a few core principles in mind and remaining flexible will allow you to eat well even when eating out.
- Try having regular meals with your family at least a few times a week, if not every night. It will nourish your body and your soul.

CHAPTER 19

PREPARE FOR THE PRESCRIPTION BY BREAKING BAD HABITS

Preparing to Detoxify

One week before you start the UltraMetabolism Prescription, prepare your body for all the goodness to come by shedding habits that interfere with your metabolism. By eliminating "toxic" items from your diet in a systematic way you will make your transition into the program simple and painless.

Eating sugar and refined high-glycemic-load carbohydrates fuels the hormones that keep your appetite out of control. Stopping it for a few weeks will change your outlook forever. Are you having a panic attack right now just thinking about giving up sugar? You are not alone, and you are probably addicted to sugar. But relax; despite your disbelief, the cravings will disappear within a few days. It is the beginning of detoxifying and re-balancing your metabolism. By doing this you won't be fighting all your urges and cravings any longer.

Eliminating trans fats will, with one quick change, leave your diet free of most processed and junk food. If you find yourself thinking there will be nothing to eat, you might need to just take the leap and trust that you won't starve to death. As you have learned, trans fats, to put it mildly, are not good for you or your metabolism.

We use caffeine to keep us awake, to compensate for lack of sleep, but it creates a false energy and ultimately creates more stress in our bodies. It is a quick adrenaline rush, then we crash—and that is when we start looking for something else to perk us up, like some sugar! Try to slowly get off it the way I recommend, and you will realize that you were more tired on the coffee than off it (but give yourself a few days to catch up on the sleep that you missed having all those triple lattes).

Alcohol is one of the nectars and sweet pleasures of life, but many of us rely on it to relax, and regular use disinhibits us around food. Ever wonder why at a restaurant you are always asked to give your drink order first and then get a bread basket? If you eat some sugar (in the form of white bread) and drink a glass of wine, you will likely order more and eat more. Taking a holiday from alcohol, besides getting rid of additional sugar calories, will help you tune in to your true appetite and prevent you from overeating.

Just eliminating sugar, refined carbohydrates, trans fats, caffeine, and alcohol can have profound effects on the way you feel and your weight in a very short time, even if you do nothing else!

Items to Eliminate

Over the course of the next week, you should eliminate these items from your diet entirely. Remember, in some cases they are hidden in places you may not expect. Be as vigilant as you can about reading labels and making sure the foods you eat do not contain the following:

- Caffeine
- Processed and refined carbohydrates (white or wheat flour) and sugar
- High-fructose corn syrup
- Hydrogenated fats and oils
- Processed and packaged foods
- Alcohol (can be reintroduced in Phase II)

How to Eliminate Caffeine in Seven Days

If you have been drinking caffeine for a long time, you will need a few days to get off it. Minimize your pain and the difficulty giving up your addiction by following these steps.

1. Begin on a weekend, when you can take naps as needed.
2. First three days: Cut down to one half your normal amount of coffee, cola, black tea, or other caffeinated beverage intake.
3. Next four days: Drink 1 cup caffeinated green tea steeped in boiling water for five minutes. You may continue green tea for all its wonderful health and weight benefits.
4. All seven days: Take 1,000 to 2,000 milligrams of buffered vitamin C powder.
5. Drink at least six to eight glasses of filtered water a day.

How to Eliminate Sugar and White Flour

Eliminating sugar is hard because it is an addiction, but the physical cravings dissipate quickly once you stop eating it. Here are some tips for how to do this successfully.

❖ Try to eliminate sugar and white flour (also known as "wheat flour") five to seven days before you begin Phase I—you will not regret it!

❖ The tried-and-true method from my experience with thousands of patients: Go cold turkey from all white flour and sugar products (don't cheat—it will only make it worse).

❖ Include protein in your breakfast, such as eggs, nuts, seeds, nut butter, tofu, or a protein shake.

❖ Combine protein, "good" fat, and carbs at each meal. Good fats are fish, extra virgin olive oil, olives, nuts, seeds, and avocados. Good carbs are beans, vegetables, whole grains, and fruit. Good proteins are fish, eggs, nuts, soy, whole grains, and legumes.

❖ Don't go low fat—consume olive oil, olives, nuts, seeds, and avocados.

❖ Eat every three hours—snack on nuts (one serving is a handful) and seeds such as almonds, walnuts, or pumpkin seeds (raw or dry-roasted only).

❖ Drink at least six to eight glasses of filtered water a day.

CHAPTER 20

PHASE I: DETOXIFY YOUR SYSTEM

The first three-week phase is designed to unlock and restart your metabolism. It is similar to rebooting your computer and getting a clear screen. Our metabolism becomes stuck in certain biochemical patterns that need to be broken. This phase is designed to break the patterns and give you a fresh start.

Breaking these patterns is step one in helping you understand exactly what has been holding you back from losing the weight that you want to lose. In Phase I, you will eliminate a lot of different things from your diet. Not only will you get rid of the poison you have been eating, you will eliminate items to which you may have allergies or sensitivities.

In Phase II you will begin to reintroduce these foods. It is only by eliminating them in the first place that you can have a realistic picture of how they have been affecting your weight and your health.

This phase often brings renewed energy, vitality, and improvement in many chronic health conditions, as well as rapid weight loss of six to eleven pounds. Some will lose significantly more, some less, but that's a good estimate.

It involves eating simple, clean, nourishing, real food. No restriction is put on the *amount* of food you eat, because counting calories, carbs, or fat grams is not the point of this program. The emphasis is on cleansing, renewal, and revitalization, and on optimizing your metabolism so it works as it was originally designed.

Three-Week Detoxification and Rejuvenation: Jump Start to Health and Weight Loss

During this three-week detoxification phase, all potentially inflammatory and allergenic foods are eliminated. This includes gluten (found mainly in wheat, rye, barley, spelt, kamut, and oats because of cross contamination), dairy, eggs, most saturated fats, and other foods I have found bothersome to my patients over the years. The diet has been developed this way so that you can start from a clean slate.

Not to fear. You will be able to reincorporate foods that you enjoy and that are healthy for you in Phase II of the prescription. You don't have to stop eating these foods forever, but it is better to eliminate them up front so that you can detoxify your body and get a real sense of what works for you.

Phase I will focus on nourishing, simple, real foods, including vegetables, fruit, legumes, nongluten grains, nuts, seeds, oils, and lean animal protein. A special nourishing detoxifying broth enhances cleansing and weight loss and can be consumed freely throughout the day.

To get started, I have provided a list of which foods you can enjoy and which foods to avoid. This is your guide for the first three weeks. The menus that follow outline the general roadmap of what you will be eating. The menus solve all the hard work of figuring out what to eat once you avoid things you may be used to eating on a regular basis. The recipes follow, giving you clear instructions (my mother always said, "if you can read, you can cook"). Finally, the shopping lists allow you to go to the store and get exactly what you need to prepare the meals.

Foods and Ingredients to Avoid in Both Phases

The following is a list of foods you will want to avoid throughout the UltraMetabolism Prescription:

- Refined or non–whole grain flour products, all types of flour: bagels, breads, rolls, wraps, pastas, etc.
- Sugar and sugar-laden foods: candy, cookies, cereals, pastries, pies, sodas, etc.
- High-fructose corn syrup
- Artificial sweeteners: aspartame (NutraSweet, Equal); saccharin (Sweet 'N Low); sucralose (Splenda); acesulfame-K (Sunette), and products/beverages with artificial sweeteners (cereals, diet drinks, etc.)
- Stevia
- Sugar alcohols: poly-ols such as mannitol, sorbitol, lactitol, xylitol, maltitol, etc.
- Artificial colors
- Hydrogenated and partially hydrogenated oils
- Canola and peanut oils
- Fat substitutes: Olean, Olestra, Salatrim, Benefat
- Unsafe additives: potassium bromate, propyl gallate, sodium nitrite, sodium nitrate, etc.
- Caffeinated beverages: sodas, coffee, tea, waters

For more information about additives in your food, check the Center for Science in the Public Interest Web site at www.cspinet.org/reports/chemcuisine.htm.

Foods to Enjoy

The following is a list of foods you can enjoy throughout the course of Phase I. You eat all of these foods over the course of the program. Please note that in some cases you will see foods on this list that are not included in the recipes. That is to give you a little room for variety. If there are foods you like better in this list than what is used in the recipes, feel free to substitute.

Fruit

Apple (all varieties)
Apricot
Avocado
Banana
Blackberries
Blueberries
Cantaloupe
Clementine
Cranberries
Figs, fresh
Grapefruit
Grapes
Honeydew melon
Kiwi
Lemon
Lime
Nectarine
Orange
Papaya
Peach
Pear
Persimmon
Pomegranate
Plum
Raspberries
Star fruit
Strawberries
Watermelon

Vegetables

Artichokes
Arugula
Asparagus
Beans, green and yellow
Beets
Bell pepper (all colors)
Bok choy
Broccoli
Brussels sprouts
Burdock root★
Cabbage (all varieties)
Carrots
Cauliflower
Celery
Celery root (celeriac)
Chile peppers

★Starchy vegetable

Vegetables *(continued)*

Collard greens

Corn★

Cucumber (pickling [Kirby] and regular)

Daikon radish and leaves

Dark leafy greens (collards, kale, spinach, Swiss chard)

Eggplant

Endive

Escarole

Fennel

Garlic

Ginger, fresh

Jicama

Kale

Kohlrabi

Leeks

Lettuce (all varieties)

Mushrooms (shiitake and other varieties)

Okra

Onions

Parsnips★

Peas★

Peppers (yellow, orange, green)

Potatoes (Red Bliss and russet)★

Pumpkin★

Radicchio lettuce

Radish (daikon and red)

Rutabaga★

Sea vegetables (arama, dulse, hijiki, wakane, etc.)

Snow peas

Spinach

Sprouts (all varieties)

Squash (summer and winter varieties)★

Sweet potatoes★

Swiss chard

Tomato

Turnips★

Water chestnuts

Watercress

Zucchini

Grains (gluten-free)

Amaranth

Brown rice

Buckwheat groats (kasha)

Corn and sprouted corn tortillas

Millet

Quinoa

Teff

Wild rice

Beans

Adzuki beans

Black beans

Chickpeas (garbanzo beans)

Great Northern beans

★Starchy vegetable

Kidney beans (red and white)
Lentils
Lima beans
Navy beans
Pinto beans
Refried beans, vegetarian
Soy milk (plain)

Soy yogurt (plain)
Soybeans and edamame
Split peas
Tofu
White beans (cannellini or
 Northern)

Fish and Seafood

The following are the best ecological choices and have the fewest contaminants. For more information on quality seafood, check Oceans Alive's website at www.oceansalive.org and Monterey Bay Aquarium Seafood Watch (www.seafoodwatch.org). To calculate your exposure to mercury from fish check www.gotmercury.org.

Abalone, U.S. farmed
Anchovies
Arctic char, U.S. and Canadian
 farmed
Catfish, U.S. farmed
Clams
Crawfish, U.S.
Halibut, Alaskan, small
Herring, Atlantic sea herring
Mackerel, Atlantic
Mussels, farmed blue, New
 Zealand green

Oysters, farmed Eastern European, Pacific
Sablefish (black cod, Alaskan)
Salmon, wild Alaskan
Sardines
Scallops, farmed bay
Shrimp, Northern from Newfoundland, U.S. farmed
Striped bass, farmed
Tilapia, U.S.

Poultry*

Chicken, skinless white breast
 meat
Cornish hen

Turkey, skinless white breast
 meat

Red Meat*

Lamb (loin)

---

* Lean, organic, grass fed/finished preferred

Nuts and Seeds

Almonds

Brazil nuts

Cashews

Coconut, fresh and unsweetened
  dried

Filberts

Flaxseeds

Hazelnuts

Hempseed

Macadamia nuts

Pecans

Pine nuts (pignoli)

Pumpkin seeds

Sesame seeds

Soy nuts

Sunflower seeds

Tahini (sesame seed paste)

Walnuts

Natural nut and seed butters
  made from these nuts and
  seeds

Fats and Oils

Almond oil

Avocado oil

Coconut butter and oil

Flaxseeds (ground and as
  flaxseed oil)

Grapeseed oil

Macadamia nut oil

Olive oil (extra virgin)

Olives, black and green

Palm fruit oil

Pumpkin seed oil

Sesame seed oil (plain and
  toasted)

Walnut oil

Beverages

Filtered water

Green tea, decaffeinated

Herbal tea

Flavorings and Seasonings*

All fresh and dried whole or
  ground herbs and spices (allspice, basil, cilantro, curry, etc.)

Broths, low-sodium organic
  (vegetable and chicken)

Chinese 5-spice powder

Chocolate, dark, minimum 70%
  cocoa

Cocoa nibs

Cocoa powder

Curry paste (red and green)

Garam masala

Herbamare (seasoning salt)

Horseradish

Kelp powder

Pepper, black and white

* Be sure to purchase gluten-free brands (refer to resource list)

Pomegranate molasses
Salsa
Sea salt

Soy sauce, low-sodium and
gluten-free
Wasabi powder

Foods to Avoid

The following is a list of foods to avoid during Phase I. You will be able to reintegrate some of these foods in later phases.

Fruit

Dried fruit
Canned fruit (syrup packed)
Fruit juices

Grains

Gluten-containing grains and grain products

Barley
Bulgur (wheat)
Couscous (wheat)
Durum (wheat)
Kamut (wheat)
Malt, malt extract, malt
  flavoring, malt
  syrup
Oats and oat bran*
Rye
Semolina (wheat)
Spelt
Sprouted wheat
Triticale
Wheat, whole wheat
Wheat berries
Wheat bran
Wheat flours (white, wheat,
  enriched wheat, stone ground

whole-wheat, 100% whole-
  wheat)
Wheat germ
Wheat starch
White rice
Food products made from the
  above grains such as:
  • Bread
  • Bread crumbs
  • Cake
  • Cereal (hot and cold)
  • Cookies
  • Crackers
  • Matzoh
  • Pasta
  • Pie
  • Rolls
  • Tortillas (wheat)
  • Wraps

---

* Oats may be cross-contaminated with gluten

Beans

Miso   Tempeh

Dairy (all dairy products)

Butter   Ice cream and ice milk
Cheese (all types)   Milk (whole, 2%, 1%, skim, non-
Cottage cheese   fat)
Cream   Sheep's milk and sheep's cheese
Cream cheese   Sour cream
Goat's milk and goat's cheese   Yogurt (cow, goat, or sheep)
Half-and-half   Frozen yogurt

Eggs

All

Poultry

Chicken (with skin on)   Turkey (with skin on)
Processed poultry products

Beef

All cuts, fresh and processed

Pork

All cuts, fresh and processed   Ham
Bacon   Sausage

Deli Meats

All processed meats
Organ meats

Nuts and Seeds

Peanuts and peanut butter   Pistachios

Fats and Oils

Canola (rapeseed) oil   Wheat germ oil
Peanut oil

Flavorings and Seasonings

Note: many packaged foods, spices, and flavorings may contain gluten. If you think you have a significant intolerance or reaction to gluten, I recommend you work with a dietician.

Seasoning mixes that are not whole or ground herbs and spices (e.g., taco seasoning, Ranch, southwest, creole, etc.)

Flavorings (may contain gluten)
Soy sauce and tamari, made from wheat

Beverages

Alcohol (all types)
Caffeinated beverages (coffee, tea, soda, water)
Chocolate drinks
Coffee substitutes

Fruit drinks and fruit punches
Instant teas
Malted beverages
Regular and diet sodas

Phase I Menus

Day 1

Breakfast: Apple-Walnut Amaranth (p. 252)
Snack: Seasonal Fresh Fruit and Raw Nuts and Seeds (p. 270)
Lunch: White Beans on a Bed of Greens (p. 256)

Snack: Avocado with Lemon (p. 271)
Dinner: Wild Salmon with Rosemary Sweet Potatoes and Lemon
   Asparagus (p. 262)

Day 2

Breakfast: Berriest Smoothie (p. 253)
Snack: Seasonal Fresh Fruit and Raw Nuts and Seeds (p. 270)
Lunch: Tarragon Chicken Salad (p. 257)
Snack: Olive Tapenade and Raw Veggies (p. 272)
Dinner: Black Bean–Cocoa Soup with Lime Zest (p. 263); Arugula
   Salad with Golden Vinaigrette (p. 264)

Day 3

Breakfast: Hot Brown Rice, Nuts, and Flax (p. 253); unsweetened
   plain soy milk
Snack: Seasonal Fresh Fruit and Raw Nuts and Seeds (p. 270)
Lunch: Avocado and Bean Burrito (p. 258)
Snack: Dark Chocolate or Cocoa Nibs and Fresh Coconut (p. 271)
Dinner: Moroccan Chicken with Cauliflower and Cashews
   (p. 265)

Day 4

Breakfast: Nut Butter Smoothie (p. 254)
Snack: Seasonal Fresh Fruit and Raw Nuts and Seeds (p. 270)
Lunch: Asian Bean Salad with Tahini Dressing (p. 258)
Snack: Artichoke Paste with Raw Veggies (p. 271)
Dinner: Coconut Dal with Steamed Broccoli and Brown Rice
   (p. 266)

Day 5

Breakfast: Peach Quinoa with Flax and Nuts (p. 254); unsweetened
   plain soy milk
Snack: Seasonal Fresh Fruit and Raw Nuts and Seeds (p. 270)
Lunch: Curried Waldorf Salad (p. 259)
Snack: Tahini with Flax Crackers (p. 271)
Dinner: Sesame-Crusted Sole with Baby Bok Choy and Wild Rice
   (p. 267)

Day 6

Breakfast: Berriest Smoothie (p. 253)
Snack: Seasonal Fresh Fruit and Raw Nuts and Seeds (p. 270)
Lunch: Cashew-Shrimp Lettuce Wraps (p. 260)
Snack: Olive Tapenade and Raw Veggies (p. 272)
Dinner: Orange Chicken with Escarole and Steamed Kasha (p. 268)

Day 7

Breakfast: Hot Buckwheat, Banana, Flax, and Walnuts (p. 255)
Snack: Seasonal Fresh Fruit and Raw Nuts and Seeds (p. 270)
Lunch: Roast Turkey Breast and Avocado Cream on a Pile of Greens
(p. 261)
Snack: Dark Chocolate or Cocoa Nibs and Fresh Coconut (p. 271)
Dinner: Lamb and Vegetable Curry with Steamed Brown Rice (p. 269)

Dr. Hyman's Detox Broth

Makes approximately 8 cups (2 quarts)
Prep time: 30 minutes
Cook time: 2 hours

The broth is a wonderful, filling snack that will also provide you with many healing nutrients. Prepare at the beginning of the week and enjoy a few cups a day.

10 cups filtered water
6 cups chopped mixed organic veggies
Fresh or dried herbs and spices such as bay leaf, oregano, lemongrass,
fennel, and ginger

Use a variety of vegetables that include at least four of the following: shiitake mushrooms, burdock root, sweet potatoes, carrots, onions, celery, sea vegetables, dark leafy greens, daikon and daikon leaf, potatoes, and other root vegetables such as parsnips, rutabagas, or turnips.

Add the herbs to the veggies and water and bring to a boil in a large stockpot. Lower the heat and simmer for a few hours.
Strain and drink warm 2 to 3 cups a day.
Keep in the refrigerator for 3 to 5 days.
Store in a tightly sealed glass container.

Phase I

Breakfast Recipes

Apple-Walnut Amaranth

Makes four ⅔-cup servings
Prep time: 5 minutes
Cook time: 30 minutes

1 cup amaranth
3 cups plain soy milk
¼ teaspoon ground cinnamon
Pinch sea salt (optional)
1 large apple, skin on, cored and diced
½ cup chopped walnuts

Place the amaranth, soy milk, cinnamon, salt (if using), and apple in a medium saucepan. Bring to a boil, stirring frequently. Cover pan and reduce heat to low. Simmer for 25 to 30 minutes until amaranth is soft. Top with chopped walnuts and serve.

Chef Tip: Rinsing and soaking grains such as amaranth for a few hours before or overnight will help reduce cooking time. To save time in the morning, you can combine all the ingredients except the walnuts in a covered saucepan the night before, store it in your refrigerator, and cook in the morning. Store leftover grain in a tighty covered glass bowl and freeze for a busy morning. Thaw the night before in the fridge. Reheat on the stove top.

SwiftTip: Amaranth, an ancient golden seed, is making a comeback because of its nutritional profile and delicious taste. It is a good source of protein, calcium, B vitamins, and fiber.

Nutrient Notables

Per serving: calories 380, carbohydrate 48g, fiber 10g, protein 16g, fat 15g, cholesterol 0mg, sodium 83mg, calcium 370mg

Note: Use only gluten-free soy milks; some brands are fortified with calcium, B12, and other nutrients, thus the vitamin and mineral content may be higher than in the analysis provided if fortified products are used.

Berriest Smoothie

Makes 1 serving
Prep time: 5 minutes
Cook time: None

½ cup plain soy milk
½ cup plain soy yogurt
1½ cups fresh or frozen mixed berries
1 tablespoon flaxseeds, ground

Place all the ingredients in a blender and mix until smooth.

SwiftTip: Flaxseeds have a mild, nutty flavor and provide a rich source of alpha-linolenic acid, an omega-3 fatty acid, dietary fiber, and lignans, unique cancer-blocking components.

Nutrient Notables

Per serving: calories 287, carbohydrate 48g, fiber 10g, protein 10g,
fat 8g, cholesterol 0mg, sodium 80mg, calcium 250mg

Hot Brown Rice, Nuts, and Flax

Makes two ¾-cup servings
Prep time: 5 minutes
Cook time: 50 minutes

½ cup long-grain brown rice
1 cup plain soy milk
¼ teaspoon ground nutmeg
Pinch sea salt (optional)
8 Brazil nuts, shelled and chopped
2 tablespoons flaxseeds, ground

Place the brown rice, soy milk, nutmeg, and salt (if using) in a medium saucepan. Bring to a boil, stirring frequently. Cover pan and reduce heat to low. Simmer for approximately 45 minutes. Top with chopped Brazil nuts and ground flaxseeds.

Nutrient Notables

Per serving: calories 372, carbohydrate 46g, fiber 6g, protein 12g,
fat 17g, cholesterol 0mg, sodium 56mg, calcium 240mg

Nut Butter Smoothie

Makes 1 serving
Prep time: 5 minutes
Cook time: None

½ cup plain soy milk
¼ cup drained silken tofu
½ small frozen banana
1.5 tablespoons natural nut butter (almond or cashew)
Ice (optional)

Place all the ingredients in a blender and mix until smooth.

Chef Tip: Peel bananas, cut in half, wrap in wax paper, freeze—and they're smoothie ready!

SwiftTip: Bananas are rich in B6 and potassium, and taste great when paired with nut butters in a smoothie or as a snack.

Nutrient Notables

Per serving: calories 295, carbohydrate 24g, fiber 4g, protein 13g,
fat 18g, cholesterol 0mg, sodium 74mg, calcium 130mg

Peach Quinoa with Flax and Nuts

Makes four ¾-cup servings
Prep time: 5 minutes
Cook time: 25 minutes

1 cup quinoa, thoroughly rinsed and drained
2 cups plain soy milk
¼ teaspoon ground allspice

Pinch sea salt (optional)

2 medium peaches, peeled, pitted, and diced, or 1½ cups frozen peaches

2 tablespoons flaxseeds, ground

2 tablespoons chopped hazelnuts

Place the quinoa, soy milk, allspice, salt (if using), and peaches in a medium saucepan. Bring to a boil, stirring frequently. Cover pan and simmer on low heat for approximately 20 minutes until quinoa is tender. Top with ground flaxseed and chopped hazelnuts.

Nutrient Notables

Per serving: calories 285, carbohydrate 41g, fiber 5g, protein 12g,
fat 9g, cholesterol 0mg, sodium 59mg, calcium 230mg

Hot Buckwheat, Banana, Flax, and Walnuts

Makes four ⅔-cup servings
Prep time: 5 minutes
Cook time: 25 minutes

1 cup buckwheat groats, whole (kasha)

2 cups plain soy milk

¼ teaspoon ground cinnamon

Pinch sea salt (optional)

1 small banana, mashed

2 tablespoons flaxseeds, ground

2 tablespoons chopped walnuts

Place the buckwheat, soy milk, cinnamon, salt (if using), and mashed banana in a medium saucepan. Bring to a boil, stirring frequently. Cover pan and reduce heat to low. Simmer for 15 to 20 minutes, until buckwheat is tender. Top with ground flaxseeds and chopped walnuts.

Nutrient Notables

Per serving: calories 270, carbohydrate 43g, fiber 7g, protein 12g,
fat 8g, cholesterol 0mg, sodium 55mg, calcium 210mg

Phase I

Lunch Recipes

White Beans on a Bed of Greens

Makes 2 servings
Prep time: 10 minutes
Cook time: None

2 cups drained canned white beans
3 tablespoons freshly squeezed lemon juice
½ cup chopped flat-leaf parsley
1 clove garlic, pressed
2 tablespoons extra virgin olive oil
Pinch sea salt
Dash freshly ground black pepper
4 cups fresh mixed baby greens

In a medium bowl, mix the beans with the lemon juice, parsley, garlic, oil, salt, and pepper. Divide the greens between two plates and spoon the white bean mixture onto the bed of greens.

Chef Tip: A citrus reamer is an excellent tool for extracting the juice from lemons and limes. Strain pits from the juice.

Swift Tip: The humble garlic bulb is nature's pungent source of the phytochemical allicin, a powerful detoxifying and antimicrobial agent.

Nutrition Notables

Per serving: calories 228, carbohydrate 32g, fiber 7g, protein 11g, fat 7g, cholesterol 0mg, sodium 56mg, calcium 130mg

Tarragon Chicken Salad

Makes 2 servings
Prep time: 15 minutes
Cook time: None

8 ounces cooked chicken breast, diced into 1-inch cubes
3 cups fresh watercress, washed and separated, thick stems removed
5 red radishes, chopped
2 stalks celery, chopped
1 medium pear, skin on, cored and diced
⅓ cup pine nuts
3 tablespoons chopped fresh tarragon, or 1 tablespoon dried
⅛ teaspoon ground cardamom
1 tablespoon walnut oil

In a large mixing bowl, toss all ingredients together. Serve.

Chef Tip: Omit the chicken, and add 1 cup drained cooked or canned beans such as chickpeas, black beans, or red beans. Toss together and serve. You can substitute any fresh or dried herb or spice such as ground fenugreek, dill, chives, cumin, etc., as desired to flavor the beans.

Swift Tip: Watercress, a member of the mustard family with a peppery flavor, available year round, is the richest natural source of a compound called phenylethylisothiocyanate (PEITC). More than fifty scientific studies have demonstrated that PEITC is a potent detoxifying agent and inhibitor of cancer cell development.

Nutrient Notables

Per serving: calories 387, carbohydrate 18g, fiber 5g, protein 32g,
fat 24g, cholesterol 68mg, sodium 136mg, calcium 110mg

Avocado and Bean Burrito

Makes 2 servings
Prep time: 15 minutes
Cook time: None

2 cups shredded romaine lettuce
2 tablespoons yellow onions, chopped
½ medium avocado, peeled, pitted, and chopped
2 tablespoons chopped cilantro
4 tablespoons chunky tomato salsa
½ cup nonfat vegetarian refried beans
2 sprouted corn tortillas

Mix the lettuce, onion, avocado, cilantro, and salsa in a medium bowl until the vegetables are evenly coated. Smear half of the beans on each tortilla, fill with vegetable mixture, and wrap burrito style.

Chef Tip: You won't miss the cheese at all in this delicious burrito loaded with fresh veggies.

Swift Tip: Cilantro, the stems and leaves of the coriander plant, adds vitamin C, carotenoids, and folic acid, along with its high dose of lively flavor.

Nutrient Notables

Per serving: calories 288, carbohydrate 35g, fiber 9g, protein 7g, fat 15g, cholesterol 0mg, sodium 379mg, calcium 90mg

Asian Bean Salad with Tahini Dressing

Makes 2 servings
Prep time: 15 minutes
Cook time: None

Tahini Dressing
¼ cup tahini
2 tablespoons extra virgin olive oil
1 tablespoon minced garlic
2 tablespoons freshly squeezed lemon juice

Pinch sea salt
Dash freshly cracked black pepper

4 cups fresh baby spinach
¼ cup chopped scallions
½ cup snow peas, strings removed
1 cup bean sprouts, rinsed and drained
1 cup drained canned adzuki beans

In a small bowl, whisk together the tahini, olive oil, garlic, lemon juice, salt, and pepper. Place the spinach, scallions, snow peas, bean sprouts, and beans in a large salad bowl. Pour the tahini dressing over the vegetable mixture and toss together to coat. Serve.

Chef Tip: Tahini is a paste made from ground sesame seeds; it is often an ingredient in hummus.

Nutrient Notables

Per serving: calories 426, carbohydrate 35g, fiber 12g, protein 13g, fat 28g, cholesterol 0mg, sodium 160mg, calcium 200mg

Curried Waldorf Salad

Makes 2 servings
Prep time: 15 minutes
Cook time: None

1 large Red Delicious or Gala apple, skin on, cored and diced
1 cup extra-firm tofu, drained well and cut into 1-inch cubes
½ cup chopped celery
¼ cup chopped toasted walnuts
½ tablespoon flaxseeds, ground
½ teaspoon grated fresh ginger
½ teaspoon curry powder
1 tablespoon walnut oil
head endive, separated and washed

In a large bowl, combine the apple, tofu, celery, walnuts, flaxseed, ginger, curry powder, and oil. Arrange endive in layers on salad plates. Spoon the apple-tofu mixture on the endive and serve.

Chef Tip: Toasting nuts brings out their rich flavor. Sprinkle nuts in a single layer on a cookie sheet and toast at 350 degrees for 10 to 15 minutes, stirring occasionally.

SwiftTip: Ginger is a knobby root with a pungent aroma and intense flavor. It adds an anti-inflammatory zip to any dish. Try fresh ginger slices in filtered water with a sprig of mint for a refreshing beverage, or warm ginger tea as a soothing evening brew.

Nutrient Notables

Per serving: calories 286, carbohydrate 28g, fiber 12g, protein 16g, fat 16g, cholesterol 0mg, sodium 88mg, calcium 350mg

Cashew-Shrimp Lettuce Wraps

Makes 2 servings
Prep time: 20 minutes
Cook time: None

Cashew Sauce
¼ cup natural cashew butter
1 tablespoon coconut milk
3 tablespoons freshly squeezed lime juice
¼ teaspoon chili powder

½ pound cooked shrimp, thawed if frozen
¼ cup chopped scallions
½ cup shredded carrots
½ cup bean sprouts, washed and drained
½ cup diced seeded cucumber
¼ cup sesame seeds
1 tablespoon grated fresh ginger or 1 teaspoon dried
½ tablespoon low-sodium, gluten-free soy sauce
2 tablespoons freshly squeezed lime juice
¼ cup unseasoned rice vinegar
6 large leaves Boston lettuce, washed

In a small bowl or blender, whip the cashew butter, coconut milk, 3 tablespoons lime juice, and chili powder together. In a large bowl, combine the shrimp, scallions, carrots, bean sprouts, cucumber, sesame seeds, ginger, soy sauce, 2 tablespoons lime juice, and vinegar. Mix well and let flavors marinate in the refrigerator for 20 minutes. Divide the cashew sauce and shrimp-vegetable mixture in center of each leaf and wrap lettuce around filling. Wrap and serve.

Chef Tip: You can use other nut butters to make this delicious sauce which adds great flavor to salads, wraps, and stir-frys.

Thaw shrimp under cold running water.

Nutrient Notables

Per serving: calories 436, carbohydrate 22g, fiber 5g, protein 33g, fat 26g, cholesterol 215mg, sodium 430mg, calcium 100mg

Roast Turkey Breast and Avocado Cream on a Pile of Greens

Makes 2 servings
Prep time: 20 minutes
Cook time: None

Avocado Cream
1 large avocado, peeled and pitted
¼ cup freshly squeezed lemon juice
3 tablespoons extra virgin olive oil
1 clove garlic, minced
Pinch sea salt
Dash freshly ground black pepper

6 cups fresh mixed baby greens
6 to 8 ounces roast skinless, sliced turkey breast
½ small red onion, sliced in slivers
1 pickling cucumber (Kirby), sliced thinly
10 pitted green olives, drained and chopped

In a food processor, blend the avocado, lemon juice, olive oil, garlic, salt, and pepper. Slowly add up to ¼ cup filtered water and process until the dressing has a creamy consistency.

Place the greens on two serving plates and top with the turkey, onion, cucumber, and olives. Drizzle with the avocado cream.

Chef Tip: This dressing can be refrigerated covered, and will remain fresh for 2 to 3 days.

SwiftTip: Avocados and olives are rich in healthy monounsaturated fats.

Nutrient Notables

Per serving: calories 304, carbohydrate 12g, fiber 6g, protein 31g,
fat 15g, cholesterol 70mg, sodium 307mg, calcium 100mg

Phase I

Dinner Recipes

Wild Salmon with Rosemary Sweet Potatoes and Lemon Asparagus

Makes 2 servings
Prep time: 20 minutes
Cook time: 25 minutes

2 small sweet potatoes
1 small yellow onion
2 tablespoons extra virgin olive oil
Pinch sea salt
1 clove garlic, pressed
2 teaspoons dry mustard
1 tablespoon freshly squeezed lemon juice
1 tablespoon chopped fresh rosemary
½ pound fresh asparagus, trimmed
Grated zest of 1 lemon
8 ounces wild salmon fillets cut into two 4-ounce portions

Preheat the oven to 425 degrees. Cut a piece of parchment paper to fit a shallow baking pan.

Wash the sweet potatoes. Slice the potatoes and onions ¼-inch thick. Put sweet potatoes and onions on the baking sheet in a single layer. Drizzle with the olive oil and sprinkle with the salt. Bake for 15 minutes.

Meanwhile, mix the garlic, mustard, lemon juice, and rosemary to make a paste and set aside.

Remove the sweet potatoes and onions from oven (keep in the baking pan). Place the asparagus on the paper around the sweet potatoes and onions. Sprinkle the lemon zest on the asparagus. Lay the salmon on top of the asparagus and onions. Spread the paste on top of the salmon.

Return the pan to oven and roast for 12 minutes. Salmon is done when the flesh flakes with gentle pressure.

Chef Tip: Parchment paper is a special nonstick paper that will not burn in the oven. Simply throw it away after use; you will not need to wash the baking pan. Do not substitute wax paper. Use only natural parchment paper.

SwiftTip: "Rosemary for remembrance," an age-old adage, holds true today as this aromatic herb in the mint family offers antioxidant and anti-inflammatory benefits important for memory.

Nutrient Notables

Per serving: calories 390, carbohydrate 22g, fiber 5g, protein 29g, fat 21g, cholesterol 50mg, sodium 77mg, calcium 120mg

Black Bean–Cocoa Soup with Lime Zest

Makes 4 servings
Prep time: 15 minutes
Cook time: 50 minutes

2 tablespoons extra virgin olive oil
1 small red onion, chopped
3 cloves garlic, pressed
1 large carrot, chopped
1 stalk celery, chopped
3 cups low-sodium organic vegetable broth
2 tablespoons cocoa powder
1 teaspoon ground cumin
2 cups canned black beans
Grated zest of 1 lime

Place the olive oil in a medium saucepan and add the onion. Sauté on low heat until onions are caramelized, about 15 minutes. Add the garlic, carrots, and celery and cook for 5 minutes. Add the vegetable stock, cocoa powder, and cumin. Stir well. Simmer for 10 minutes. Stir in the black beans. Add the lime zest. Cook for approximately 20 minutes over low heat.

Serve with Arugula Salad with Golden Vinaigrette.

Chef Tip: Experiment using cocoa powder in some of your favorite recipes such as baked beans, soups, and stews for a whole new flavor experience.

Swift Tip: Cocoa powder and chocolate, aptly known as food of the gods, is produced from the plant *Theobroma cacao*. It is recognized for its super antioxidant polyphenolic compounds and the friendly anti-inflammatory fat, oleoethanolamide (OEA).

Nutrient Notables

Per 1-cup serving: calories 248, carbohydrate 37g, fiber 11g, protein 10g, fat 8g, cholesterol 0mg, sodium 128mg, calcium 120mg

Arugula Salad with Golden Vinaigrette

Makes nine 2-tablespoon servings vinaigrette
Prep time: 10 minutes
Cook Time: None

¾ cup extra virgin olive oil
¼ cup freshly squeezed lemon juice
2 tablespoons Dijon mustard
1 teaspoon ground turmeric
Pinch sea salt
Dash freshly ground black pepper
4 cups trimmed arugula, washed

In a small bowl, whisk together the oil, juice, mustard, turmeric, salt, and pepper. Drizzle over fresh arugula. Toss to coat.

Chef Tip: This dressing can be refrigerated up to 10 days.

Nutrient Notables

Dressing, per 2-tablespoon serving: calories 184, carbohydrate 1g, fiber 0g, protein 0g, fat 20g, cholesterol 0mg, sodium 43g, calcium 0mg
Arugula, per serving: calories 50, carbohydrate 5g, fiber 2g, protein 2g, fat 0g, cholesterol 0mg, sodium 30mg, calcium 40mg

Moroccan Chicken with Cauliflower and Cashews

Makes 4 servings
Prep time: 30 minutes
Cook time: 30 minutes

1 tablespoon extra virgin olive oil
½ pound skinless, boneless chicken breasts, cut into 1-inch cubes
1 small yellow onion, chopped
3 cloves garlic, pressed
2 cups cauliflower florets
2 cups drained canned chickpeas
6 cups low-sodium organic chicken broth
2 tablespoons pomegranate molasses
1 tablespoon garam masala
½ cup raw cashews, chopped

Preheat the oven to 350 degrees.

In a flameproof casserole, heat the olive oil on medium-high heat. Add the chicken. Let brown for 5 minutes, stirring occasionally. Add the onions, garlic, cauliflower, chickpeas, chicken broth, pomegranate molasses, and garam masala. Stir to mix. Cover and place in oven for approximately 25 minutes.

Top with chopped cashews before serving.

Chef Tip: The flavor of dried herbs is enhanced when added at the beginning of a recipe. Fresh herbs, however, should be added at the end of the recipe for maximum flavor.

SwiftTip: Garam masala is a ground blend of spices that may include cloves, coriander, cumin, cardamom, fennel, mace, black pepper, and nutmeg—it is a smorgasbord of phytonutrients.

Nutrient Notables

Per serving: calories 442, carbohydrate 39g, fiber 7g, protein 29g, fat 21g, cholesterol 38mg, sodium 173mg, calcium 100mg

Coconut Dal with Steamed Broccoli and Brown Rice

Makes 6 servings
Prep time: 10 minutes
Cook time: 30 minutes

2 cups yellow split peas
One 14-ounce can lite coconut milk
4 cups low-sodium vegetable broth
1 small yellow onion, sliced
3 cloves garlic, pressed
1 tablespoon grated fresh ginger
2 teaspoons ground turmeric
1 teaspoon sea salt
4 tablespoons chopped fresh cilantro
1 medium bunch broccoli, trimmed and steamed
1½ cups raw steamed brown rice

Rinse the split peas. In a large saucepan, place the split peas, coconut milk, vegetable broth, onion, garlic, ginger, turmeric, and salt. Simmer over medium heat until peas are soft, approximately 30 minutes. Sprinkle chopped fresh cilantro on top.

Serve with steamed broccoli and brown rice.

Chef Tips: When using dried peas and beans, be sure to sift through and remove any stones prior to use.

This recipe can easily be doubled and frozen for later use as a convenient lunch or dinner.

Swift Tip: Ancient healers have long used turmeric as an anti-inflammatory remedy. It is this spice that gives ballpark mustard and curry dishes their characteristic yellow color.

Nutrient Notables

Coconut dal, per 1-cup serving: calories 317, carbohydrate 51g, fiber 18g, protein 18g, fat 6g, cholesterol 0mg, sodium 490mg, calcium 90mg
Broccoli, per 1-cup serving: calories 60, carbohydrate 12g, fiber 6g, protein 4g, fat 0g, cholesterol 0mg, sodium 32mg, calcium 30 mg
Steamed brown rice per ½-cup serving: calories 110, carbohydrate 23g, fiber 2g, protein 2g, fat 1g, cholesterol 0mg, sodium 1mg, calcium 10mg

Sesame-Crusted Sole with Baby Bok Choy and Wild Rice

Makes 2 servings
Prep time: 15 minutes
Cook time: 55 minutes

½ cup wild rice
½ pound baby bok choy, trimmed
¼ cup sesame seeds
1 tablespoon sesame oil
Two 4-ounce fillets fresh sole
2 cloves garlic, pressed
2 tablespoons grated fresh ginger
Pinch sea salt
Dash freshly ground black pepper

Rinse the wild rice and place it in a medium saucepan with 1½ cups filtered water. Cover and bring to a boil. Reduce the heat to low and steam approximately 50 minutes, until the water has been absorbed and the grains have split open. Meanwhile, cut the ends off the bok choy, wash well, and set aside.

Place the sesame seeds on a plate. Lightly coat the sole with 1 teaspoon of the sesame oil. Press sole onto the sesame seeds to form a crust. Set aside.

About 10 minutes before the wild rice is done, heat a large skillet over medium-high heat until hot. Add the remaining 2 teaspoons of the sesame oil and swirl skillet to distribute evenly over the bottom. Carefully place sole in the skillet. Cook the fish for 2 to 3 minutes, leaving it undisturbed to ensure a golden brown, crunchy crust. Using a spatula, carefully turn the sole over and brown on the other side for 2 to 3 minutes. Check the fish for doneness: It should flake apart with gentle pressure when done. Remove the sole from the pan and set aside to keep warm.

Add the bok choy, garlic, and ginger to the skillet. Toss well until the bok choy begins to wilt. Sprinkle with the salt and pepper. Place the bok choy and steamed wild rice on plates and serve the fish on top.

Chef Tips: A Microplane is an excellent tool for grating ginger. It is a very sharp grater that comes in various sizes. Simply wash the piece of fresh ginger with warm water. Rinse well and pat dry. When using a Microplane, it is not necessary to peel the skin off of the ginger.

A fish spatula is a very thin, wide spatula with slats, for lifting and turning delicate items.

SwiftTip: Wild rice is actually a marsh grass with a chewy texture. It is a good source of B vitamins, iron, magnesium, zinc, and fiber. It is often used as a gluten-free "grain option" along with quinoa, brown rice, millet, amaranth, buckwheat groats (kasha), and teff.

Nutrient Notables:

Per serving: calories 420, carbohydrate 39g, fiber 7g, protein 32g, fat 16g, cholesterol 54mg, sodium 224mg, calcium 180mg

Orange Chicken with Escarole and Steamed Kasha

Makes 2 servings
Prep time: 15 minutes
Cook time: 30 minutes

3 cups low-sodium organic chicken broth
2 tablespoons freshly squeezed orange juice
Grated zest of 1 orange
2 tablespoons Dijon mustard
¼ teaspoon ground allspice
5 fresh black or green figs, diced
Two 4-ounce skinless, boneless chicken breasts
⅓ cup whole buckwheat groats (kasha)
4 cups chopped fresh escarole, stems removed, washed thoroughly and
    drained lightly (or other dark leafy greens, if you wish)

Preheat the oven to 375 degrees.

Combine 2 cups broth, the orange juice, orange zest, mustard, allspice, and figs in a blender and puree until smooth. Place the chicken in a baking dish and coat with fig sauce. Bake for 30 minutes.

Meanwhile, in a small saucepan, stir together the buckwheat and the remaining 1 cup chicken broth. Bring to a boil and simmer covered for 15 to 20 minutes, until tender. Set aside to keep warm.

Steam the escarole in a covered saucepan for approximately 3 to 5 minutes. Drain well.

Place the escarole on serving plates and top with the baked chicken. Serve with the steamed kasha.

SwiftTip: Buckwheat (kasha) is a highly nutritious seed from the same family as rhubarb and is not related to wheat. This delicious grain adds fiber, flavonoids, and variety to a gluten-free pantry.

Nutrient Notables

Orange Chicken, per 3-oz. serving: calories 233, carbohydrate 27g, fiber 4g, protein 27g, fat 4g, cholesterol 65mg, sodium 243mg, calcium 70mg

Buckwheat, ⅔-cup serving: calories 113, carbohydrate 22g, fiber 3g, protein 6g, fat 1g, cholesterol 0mg, sodium 40mg, calcium 10mg

Escarole, per 1½-cup serving: calories 53, carbohydrate 10g, fiber 5g, protein 5g, fat 5g, cholesterol 0mg, sodium 86mg, calcium 120mg

Lamb and Vegetable Curry with Steamed Brown Rice

Makes 4 servings
Prep time: 20 minutes
Cook time: 50 minutes

1 cup brown rice
2 teaspoons sesame oil
1 medium yellow onion, sliced
1 medium red bell pepper, cut into ½-inch-wide strips
2 cloves garlic, pressed
1 tablespoon grated fresh ginger
¼ teaspoon red curry paste (or more to taste)
1 pound boneless lamb loin, cut into 1-inch strips
1 cup low-sodium organic vegetable broth
1 cup lite coconut milk
1 tablespoon reduced-sodium gluten-free soy sauce
2 cups cauliflower florets
4 cups spinach, rinsed and drained well
1 tablespoon freshly squeezed lime juice
¼ cup chopped cilantro

Combine the brown rice and 2 cups filtered water in a small saucepan. Cover and bring to a boil. Reduce the heat to low and steam for approximately 45 minutes, until the water has been absorbed and the rice is tender.

When the rice has been cooking for 25 minutes, heat the oil in a large skillet over medium-high heat. Add the onion and pepper and cook until vegetables begin to soften. Add the garlic, ginger, and curry paste and mix well. Add the lamb and cook, stirring often, for approximately 3 minutes. Stir in the broth, coconut milk, and soy sauce and bring to a simmer. Add the cauliflower and reduce heat to medium-low, stirring occasionally, until lamb is cooked through and cauliflower is tender, about 10 minutes. Toss in the spinach and lime juice, and cook just until the spinach is wilted. Serve immediately over the steamed brown rice and top with chopped cilantro.

Chef Tip: Red curry paste adds a convenient zip to your recipes. It is a blend of chile peppers, garlic, lemongrass, and galanga (a rhizome with a ginger-like flavor).

Swift Tip: Spices such as chile peppers have been shown to perk up your metabolism. They are good for gut health, too!

Nutrient Notables

Lamb and vegetable curry, per 1½-cup serving: calories 273, carbohydrate 14g, fiber 4g, protein 27g, fat 13g, cholesterol 72mg, sodium 311mg, calcium 90mg

Steamed brown rice, per ½-cup serving: calories 110, carbohydrate 23g, fiber 2g, protein 2g, fat 1g, cholesterol 0mg, sodium 1mg, calcium 10mg

---

Phase I

Snacks

Seasonal Fresh Fruit and Raw Nuts and Seeds

1 medium piece or 1 cup fruit
¼ cup nuts and/or seeds

Enjoy daily. Use a variety of nuts, seeds, and fresh fruits in season.

Nutrient Notables

Fresh fruit, per 1 medium piece or 1-cup serving: calories 75, carbohydrate 20mg, fiber 3g, protein 2g, fat 0g, cholesterol 0mg, sodium 5mg, calcium varies

Nuts/seeds, per ¼-cup serving: calories 200, carbohydrate 7g, fiber 2g, protein 8g, fat 18g, cholesterol 0mg, sodium 5mg, calcium varies

Avocado with Lemon

Peel an avocado, remove the pit and the peel, cut in wedges, and splash with some freshly squeezed lemon juice.

Nutrient Notables

Avocado with lemon, per one-avocado serving: calories 322, carbohydrate 17g, fiber 13g, protein 4g, fat 29g, cholesterol 0mg, sodium 14mg, calcium 20mg

Artichoke Paste with Raw Veggies

Drain one 14-ounce can or 12-ounce jar of artichoke hearts and puree in a food processor with 1 teaspoon extra virgin olive oil and 1 teaspoon dried Italian herbs.

Nutrient Notables

Raw veggies, per 1-cup serving: calories 40, carbohydrate 7g, fiber 3g, protein 2g, fat 0g, cholesterol 0mg, sodium 20mg, calcium 50mg

Artichoke spread, per ½-cup serving: calories 70, carbohydrate 12g, fiber 5g, protein 3g, fat 3g, cholesterol 0mg, sodium 134mg, calcium 55mg

Tahini with Flax Crackers

Add 2 teaspoons lemon juice to 1 tablespoon tahini and combine.

Tahini (ground sesame seeds) is a paste that spreads like peanut butter and can be used as an ingredient in sauces, marinades, bean spreads, and more. Flax crackers add omega-3s and a delightful crunch.

Nutrient Notables

Tahini and flax crackers, per 1 tablespoon with 15-cracker serving: calories 227, carbohydrate 26g, fiber 5g, protein 5g, fat 13g, cholesterol 0mg, sodium 170mg, calcium 80mg

Dark Chocolate or Cocoa Nibs and Fresh Coconut

Choose a chocolate that is at least 70% cocoa. Or try cocoa nibs—roasted cocoa beans separated from their husks and broken into small bits. Cocoa nibs are the essence of chocolate and can be added to savory dishes and baked goods. The combination of cocoa nibs and fresh coconut (available in your produce section) makes a crunchy snack with just the right dose of sweetness.

Nutrient Notables

Dark chocolate, per 1-ounce serving: calories 149, carbohydrate 16g, fiber 2g, protein 2g, fat 9g, cholesterol 0mg, sodium 2mg, calcium 8mg

Cocoa nibs, per 2-tablespoon serving: calories 140, carbohydrate 8g, fiber 2g, protein 2g, fat 10g, cholesterol 0mg, sodium 2mg, calcium 8mg

Fresh coconut piece, per 2-inch by 2-inch by ½-inch serving: calories 160, carbohydrate 7g, fiber 4g, protein 1.5g, fat 15g, cholesterol 0mg, sodium 9mg, calcium 10mg

Olive Tapenade and Raw Veggies

Makes 6 servings
Prep time: 15 minutes
Cook time: 30 to 45 minutes

Enjoy the flavor of olives with the crunch of raw veggies. Commercial olive spreads are also available.

2 to 3 whole heads garlic, unpeeled
½ cup pitted kalamata olives
½ cup drained pitted green olives
1 tablespoon freshly squeezed lemon juice
1 tablespoon extra virgin olive oil

Preheat the oven to 350 degrees. Place the whole garlic heads on a parchment paper–lined baking sheet. Bake for 30 to 45 minutes until light brown and soft. Let cool. Separate into cloves and squeeze out of the soft flesh. Measure ½ cup roasted garlic cloves.

In the food processor, puree the garlic, olives, juice, and oil until smooth. Serve with raw veggies, such as celery, red pepper, or Daikon radish sticks, for dipping.

Nutrient Notables

Olive tapenade, per ¼-cup serving: calories 106, carbohydrate 8g, fiber 2g, protein 2g, fat 8g, sodium 490mg, calcium 70mg

Raw vegetables, per 1-cup serving: see p. 302

Phase I Shopping List

These are the total amounts for one week, including fruit and veggie snacks. It is recommended that you shop for fresh produce at least twice weekly. Feel free to make produce substitutions based on season and availability.

Vegetables

Arugula, ½ pound
Asparagus, 1 bunch
   (about ½ pound)
Baby bok choy, ½ pound
Mixed baby greens, 2 pounds
Baby spinach, ½ pound
Bean sprouts, 2 ounces
Belgian endive, 1 head
Boston lettuce, 1 head
Broccoli, 1 head
Carrots, 2 & 1 package organic
   baby carrots for snacks
Cauliflower, 1 small head or
   1½ pounds
Celery, 1 bunch
Cucumbers, 4
Daikon radish, 1

Escarole, 1 small head (½ pound)
Garlic, 5 bulbs
Ginger, fresh, 3-inch piece
Herbs (bunches): Rosemary (1),
   Parsley, flat-leaf (1), Cilantro
   (1), Tarragon (1)
Olives: 20 green, 10 kalamata
Onions: 5 yellow, 2 red
Peppers, red bell, 2
Radishes, 1 small bunch
Romaine lettuce, 1 small head
Scallions, 1 bunch
Snow peas, 2 ounces
Spinach, ½ pound
Sweet potatoes, 2
Watercress, 2 bunches

Fruit

Apples, 3
Avocados, 3
Bananas, 2
Berries, any type, 3 pints
Coconut, 1 wedge (2" x 2" x ½")
Figs, 5 (green or black)
Lemons, 4

Limes, 3
Oranges, 2
Peaches, 2 (or one 10-ounce
   package frozen if not in
   season)
Pears, 2

Nuts and Seeds

(Purchase raw nuts and seeds and store in refrigerator; these amounts include nuts/seeds for snacks.)

Almonds, ½ pound
Brazil nuts, ¼ pound
Cashews, ¼ pound
Flaxseeds, 1-pound package

Hazelnuts, ¼ pound
Pine nuts
Sesame seeds, ½ pound
Walnuts, ½ pound

Meat, Fish, and Poultry

Boneless, skinless chicken breast,
16 ounces
Boneless, skinless split chicken
breast, two, 4 ounces each

Roast turkey breast, 6 to 8 ounces
Boneless lamb loin, 1 pound
Wild salmon, 8 ounces
Fresh sole, 8 ounces

Frozen

Frozen shrimp, cooked, large
size, ½ pound

Frozen mixed berries,
one 10-ounce package

Organic Soy Products

(Found in produce and refrigerated sections)
Plain, unsweetened soy milk,
1 gallon (gluten-free brand
such as Silk)

Plain, unsweetened soy yogurt,
one 8-ounce carton (Nancy's)
Silken tofu, 1 package
Extra-firm tofu, 1 package

Whole Grains (gluten-free)

These grains will be used in all Phases.

Amaranth, 1-pound package
Brown rice, long grain, 1-pound
package
Buckwheat groats (kasha),
1-pound package

Corn tortillas (sprouted grain),
1 package
Flax crackers, 1 package
Quinoa, 1½ pounds
Wild rice, one 8-ounce box

Pantry

Artichoke hearts, 1 jar or prepared Artichoke Antipasto
(natural food brand)
Beans (canned): aduki, 1 can;

black beans, 1 can; chickpeas
(garbanzos), 1 can; white
beans, 1 can; vegetarian refried beans (nonfat), 1 can

Chocolate, dark, 1 ounce

Cocoa powder, 1 container

Coconut milk, lite, 2 cans

Yellow split peas, 2 cups or
1 package

Staples

These items will be used in both phases, and they will not be included in the Phase II shopping list since they will now be in your pantry.

Chicken stock/broth, organic,
low-sodium, and gluten-free,
three 32-ounce containers

Cocoa nibs, 1 cup

Dijon mustard, 1 jar

Nut butters: almond and cashew,
1 jar each

Oils (purchase expeller pressed
in small bottles and store in a
cool, dark place)
- Extra virgin olive oil,
1 small bottle
- Sesame and toasted sesame
oil, 1 small bottle each
- Walnut oil, 1 small bottle

Pomegranate molasses, 1 bottle

Red curry paste, 1 small jar

Rice vinegar, unseasoned,
1 bottle

Salsa, chunky style, 1 small jar

Soy sauce, reduced-sodium
gluten-free, 1 bottle

Tahini (sesame seed paste),
1 small jar

Vegetable broth, organic, low-
sodium and gluten-free, or
two 32-ounce containers

Herbs, Spices, and Seasonings

Purchase in small quantities (.25 ounce or less) as these are used in various recipes throughout all phases and others are added to this list in Phase II.

Allspice, ground

Cardamom, ground

Chili powder, ground

Cinnamon, ground

Cumin, ground

Curry powder

Dry mustard

Garam masala

Ginger, ground

Italian seasoning

Nutmeg, ground

Peppercorns, whole black

Sea salt

Tarragon, ground

Turmeric, ground

CHAPTER 21

PHASE II: REBALANCE YOUR METABOLISM AND MAINTAIN A LIFELONG HEALTHY METABOLISM

This is your roadmap to health, balance, and optimal weight. During Phase II you will reeducate your metabolism, rebalance the hormones and molecules that control your appetite, calm your nervous system, reduce inflammation and oxidative stress ("rust"), and enhance detoxification. Many lose an additional one to three pounds a week during this phase until they reach their optimal weight.

Eating an organic diet (whenever possible) of real food rich in antioxidants and detoxifying and anti-inflammatory messages for your genes is integrated into the meal plans and recipes. There will be *no counting calories, no specific ratios of macronutrients, and no calculating fat grams or carbohydrates* to worry about (although the nutrient density of these foods is provided for you in the Nutrient Notables in each recipe). The main focus is on foods that enhance your metabolism and delight your palate. Quick options and suggestions are included for those with busy lifestyles.

This is the "meat" of the program—reeducating your metabolism to allow your body to regain health (with weight loss an effortless side effect). Part of my secret is that I never suggest people lose weight. I simply educate them about how and why their metabolism doesn't work, and how to fix it. I give them the tools to use the 7 Keys to UltraMetabolism.

Once everything starts working more normally, the weight just comes off without your having to focus on it. I never focus on the pounds, just the problems! Helping people fix the problems with their metabolism is why I have written this book. The menus and recipes in this section get you well along the road to fixing your metabolism for the long term.

Reintroducing Foods

During this phase, the diet is more liberal and relaxed, and certain foods are reintroduced. You need to be aware of what is going on in your body during this time. During Phase I of the program, you may have eliminated foods you were unknowingly sensitive or even allergic to.

The reintroduction of foods to which you may be sensitive or allergic can trigger many symptoms including:

- nasal congestion
- chest congestion
- headaches
- brain fog
- joint aches
- muscle aches
- pain
- fatigue
- changes in your skin (acne)
- changes in digestion or bowel function

Ideally you should separate the introduction of the top three major allergens by two to three days each. As you reintroduce gluten, dairy, eggs, and other foods in Phase II, monitor your response carefully and avoid for twelve weeks those foods that trigger any of these symptoms. Then try them again after twelve weeks and see if you have any reactions. Be aware that reactions can be delayed for up to forty-eight hours after eating. If you still react you should stay off them for the long term, or see a physician, dietitian, or nutritionist skilled in managing food allergies.

One thing that can help you become more aware of what foods may be affecting your body is keeping a food journal. Use it to notice the connection between food and physical symptoms such as headaches, sinus congestion, fatigue, or digestive problems. Notice the connection between food and mood.

Keeping an energy and food log helps create the connection between your daily food choices, activities, stress management, and how you feel. A journal will record your continuing sense of well-being, energy, and weight loss. Having a written record of this nature is a more practical and accessible way to keep track of what is going on for you than simply trying to remember it. I've created a simple journal that you can download by going to www.ultrametabolism.com/guide.

Maintain Lifelong Healthy Metabolism

This phase is where you develop a lifelong prescription for achieving and maintaining your optimal weight. The initial four weeks of the Ultra-Metabolism Prescription—the preparation and detox phases—lead to fairly rapid weight loss for most people. New behaviors and your new relationship to food and stress will become second nature, and your daily choices about what, when, and where to eat will become obvious because of the way you feel.

You will maintain a healthy metabolism for life by eating foods that are rich in healthy fats; slowly released, low-glycemic-load, high-fiber carbohydrates; plant proteins and minimal lean animal products; as well as disease- and weight-fighting anti-inflammatory and antioxidant chemicals and phytonutrients. These foods will speak to your genes in new ways and your genes will say yes.

Learn What Works for You

Over the course of the diet, pay attention to your body and find out what works for you. You may be more like the Greenland Inuit and feel satisfied with more healthful fats in your diet, or perhaps you are more like the Pima Indians and find that more fiber-rich carbohydrates work better for you. The same result can be obtained as long as you follow the basic nutritional wisdom that is the foundation of the UltraMetabolism Prescription.

Committing to this diet for four weeks in Phase 2, these principles will become second nature to you. Choose quality, whole, real, unprocessed foods from a variety of plant sources and moderate amounts of animal protein, and the rest will follow.

Be Flexible

Within the parameters there is almost infinite flexibility—Asian, Mediterranean, traditional Mexican, and Middle Eastern cooking. But learning how to eat real food is a discovery process. Because these foods have more fiber and nutrients, they are more filling and satisfying. You can adjust the portion sizes to your needs. If you are a 350-pound man trying to lose weight, you will need much more food in a day than if you are a 150-pound woman trying to lose ten pounds.

Once you have reset your metabolism and established a pattern of eating and self-care that nourishes you, flexibility is important. Eating any food with great relish and delight—even a decadent chocolate cake or the richest ice cream—is good for the soul and the senses. While some people may have trigger foods, like the first drink for an alcoholic, most of us can enjoy treats once in a while, which is why I have included some of my favorite desserts in the recipes section, including three with chocolate for you to enjoy occasionally. Limit desserts to once or twice a week. You may enjoy one to two squares of dark chocolate a day if you love chocolate. Your body will tell you what feels good and what doesn't.

You will be drawn to those foods that feel good to your body, and from time to time that may include almost anything. Staying in balance and finding a rhythm is the key to lifelong health and a healthy metabolism. Moderation in all things is still great advice, and that includes moderation in moderation. Enjoy yourself.

I wish you all endless health and happiness.

Foods to Enjoy

Continue to enjoy a wide variety of whole, real, unprocessed organic foods. As in Phase I, here you will find the foods to enjoy and foods to avoid, the menus, recipes, and shopping lists to guide you.

The following is a list of foods that you can start to enjoy in Phase II. You will notice that some of them were on the Foods to Avoid list in Phase I. It is okay to reintegrate some of these foods now, but watch your reactions carefully. If warning signs come up telling you that you may be allergic to one of the foods you are reintroducing, you may want to eliminate it from your diet for a longer period of time or altogether.

Of course, you should also keep in mind that all the foods you enjoyed in Phase I still apply.

Fruit

Dried fruit (sulfite free)
Pure, unsweetened fruit juices used in cooking

Grains

Gluten-containing whole grains:

- Barley, whole or unhulled
- Bulgur (wheat)
- Kamut (wheat)
- Oats (steel-cut) and oat bran
- Rye, whole
- Spelt (wheat)
- Sprouted whole wheat
- Triticale
- Wheat, whole wheat
- Wheat berries
- Wheat bran
- Wheat germ, raw

Food products made from the above 100% whole or sprouted

Beans

Miso   Tempeh

Red Meat (grass fed/finished beef) and Pork

Tenderloin, trimmed   Sirloin, trimmed

Nuts and Seeds

Peanuts and peanut butter   Pistachios

Fats and Oils

Almond oil   Olives, black and green
Avocado oil   Palm fruit oil
Coconut butter and oil   Pumpkin seed oil
Flaxseeds (ground and as   Sesame seed oil (plain and
    flaxseed oil)   toasted)
Grapeseed oil   Walnut oil
Macadamia nut oil   Wheat germ oil
Olive oil (extra virgin)

Dairy (preferred types)
(Choose organic dairy products)

Cheese (all types unprocessed) Sheep's milk, sheep's cheese,
Cottage cheese   sheep's yogurt
Cream cheese   Sour cream
Goat's milk, goat's cheese, goat's Yogurt (cow, goat, sheep)
    yogurt
Milk, 1%, skim/nonfat

Eggs

(Omega-3 type)

Sweeteners

Agave nectar   Honey

Alcohol

If you enjoy alcohol, you may reintroduce it but limit to one drink, three to four times a week. The following are good measures for what constitutes one drink:

- ✣ 4 to 5 fluid ounces wine
- ✣ 1.5 fluid ounces spirits
- ✣ 12 fluid ounces beer or 16 fluid ounces light beer

Foods to Avoid

The following are foods you will want to avoid in Phase II. Some of these overlap with Phase I. Remember to continue avoiding the foods you want to eliminate from your diet in every phase of the program.

Grains

White rice
Grain products that are not 100% whole or sprouted grain
White flour products

Poultry

Poultry with skin on   Processed poultry products

Red Meat

Grain fed

Pork

Bacon   Sausage
Ham

Deli Meats

All processed meats

Liver and Organ Meats

Beverages

Caffeinated beverages (coffee,   Instant teas
   teas, sodas, waters)   Sodas
Fruit drinks and fruit punches

Phase II Menus

Day 1

Breakfast: Whole Rye Toast with Almond Butter and Banana (p. 284)
Snack: Seasonal Fresh Fruit and Raw Nuts and Seeds (p. 270)
Lunch: Dilled Egg Salad on Baby Spinach (p. 288)
Snack: Avocado with Lemon (p. 271)
Dinner: Pan-Seared Scallops with Peach-Kiwi Salsa, Baby Bok Choy, and Steamed Kasha (p. 294)

Day 2

Breakfast: Sprouted Grain Cereal with Berries and Soy Milk (p. 284)
Snack: Seasonal Fresh Fruit and Raw Nuts and Seeds (p. 270)
Lunch: Greek Salad Pita (p. 289)
Snack: Yogurt with Fresh Fruit and Shredded Coconut and Wheat Germ (p. 303)
Dinner: Blackstrap Vegetarian Chili with Baby Greens Salad and Citrus Vinaigrette (p. 295)

Day 3

Breakfast: Vegetable-Egg Scramble with Whole-Grain Flax Toast and Fresh Fruit Medley (p. 285)
Snack: Seasonal Fresh Fruit and Raw Nuts and Seeds (p. 270)

Lunch: Wild Salmon Cakes with Asian Cucumber Cabbage Slaw (p. 290)
Snack: UltraMetabolism Road Mix (p. 304)
Dinner: Raspberry-Pistachio Crusted Chicken with Steamed Kale and Wild Rice (p. 296)

Day 4

Breakfast: Soy-Nut Pancakes with Strawberry-Banana Sauce (p. 286)
Snack: Seasonal Fresh Fruit and Raw Nuts and Seeds (p. 270)
Lunch: Black Bean Confetti Salad (p. 291)
Snack: Brazil Nut Bars (p. 305)
Dinner: Mustard-Crusted Lamb Chops with Garlicky Broccoli Rabe (p. 298)

Day 5

Breakfast: Gingered Yogurt Berry Parfait (p. 287)
Snack: Seasonal Fresh Fruit and Raw Nuts and Seeds (p. 270)
Lunch: Quinoa Tabouleh Salad (p. 292)
Snack: Dark Chocolate Square
Dinner: Cajun-Spiked Halibut with Peppers, Mustard Greens, and Baked Sweet Potatoes (p. 299)

Day 6

Breakfast: Steel-Cut Cinnamon-Apple Oats with Flax and Nuts (p. 287)
Snack: Seasonal Fresh Fruit and Raw Nuts and Seeds (p. 270)
Lunch: Wasabi Salmon Salad (p. 293)
Snack: Antipasto (p. 303)
Dinner: Chicken-Vegetable Fajitas in Sprouted Grain Tortillas (p. 300)

Day 7

Breakfast: Berriest Smoothie (p. 253)
Snack: Seasonal Fresh Fruit and Raw Nuts and Seeds (p. 270)
Lunch: Curried Turkey Salad on Watercress (p. 293)
Snack: Almond Butter with Flax Crackers (p. 304)
Dinner: Dr. Hyman's Chinese Eggs and Seasoned Greens (p. 301)

Breakfast Recipes

Whole Rye Toast with Almond Butter and Banana

Makes 1 serving
Prep time: 3 minutes
Cook time: 1 minute (toaster)

2 slices whole rye bread
1 tablespoon almond butter
1 small banana, cut into ¼-inch slices

Toast bread and spread with almond butter and banana slices.

Nutrient Notables

Per serving: calories 350, carbohydrate 56g, fiber 7g, protein 10g,
fat 12g, cholesterol 0mg, sodium 214mg, calcium 110mg

Sprouted Grain Cereal with Berries and Soy Milk

Makes 1 serving
Prep time: under 5 minutes
Cook time: None

¾ cup high-fiber, low-sugar sprouted grain cereal *
¾ cup plain soy milk
½ cup fresh berries

Place cereal, soy milk, and berries in a bowl. Enjoy!

Nutrient Notables

Per serving: calories 225, carbohydrate 40g, fiber 10g, protein 8g,
fat 4g, cholesterol 0mg, sodium 294mg, calcium 235mg

---

* Try Food for Life Ezekiel 4:9 cereal, which comes in a variety of sprouted grains.

Vegetable-Egg Scramble with Whole-Grain Flax Toast and Fresh Fruit Medley

Makes 1 serving
Prep time: 10 minutes
Cook time: 10 minutes

2 whole omega-3 eggs
1 tablespoon filtered water
1 teaspoon olive oil
1 cup assorted chopped raw vegetables (such as onions, red bell peppers, tomatoes, broccoli, zucchini, summer squash, asparagus, mushrooms)
Pinch sea salt (optional)
Dash freshly ground black pepper (optional)
2 tablespoons chunky tomato salsa
1 slice whole-grain flax toast
1 cup mixed seasonal fresh fruit (such as berries, peaches, kiwis, melon)

In a small mixing bowl, whisk the eggs and water. Heat an 8-inch skillet over medium heat. Add the olive oil and swirl pan to evenly distribute the oil. Add the vegetables and sauté for 5 minutes until they are crisp-tender. Pour the eggs over the vegetables and cook, stirring constantly, until the eggs are set. Sprinkle with salt and pepper if desired, and top with chunky tomato salsa.

Serve with whole-grain flax toast and fresh fruit in season.

Chef Tip: Frozen chopped mixed vegetables can be substituted for the fresh. Simply thaw in the refrigerator overnight and they are ready to go.

Swift Tip: Omega-3 enriched eggs are a "functional food" readily available at your local supermarket. They come from chickens fed a diet rich in algae or flaxseeds as the original source of these healthy fats. Remember, "You are what they ate."

Nutrient Notables

Eggs, per serving: calories 222, carbohydrate 8g, fiber 3g, protein 16g, fat 15g, cholesterol 423mg, sodium 194mg, calcium 90mg
Whole-grain flax toast, 1 slice: calories 90, carbohydrate 15g, fiber 3g, protein 3g, fat 1g, cholesterol 0mg, sodium 140mg, calcium 30mg
Seasonal fresh fruit medley, 1 cup: calories 75, carbohydrate 20g, fiber 3g, protein 2g, fat 0g, cholesterol 0mg, sodium 5mg, calcium 15mg

Soy-Nut Pancakes with Strawberry-Banana Sauce

Makes 4 servings, 3 pancakes each
Prep time: 15 minutes
Cook time: 5 to 7 minutes

1 small banana
2 cups fresh strawberries (or frozen unsweetened, thawed, with juice)
1 teaspoon honey
½ cup drained silken tofu
½ cup plain soy milk
2 tablespoons ground flaxseeds
¾ cup almond flour
½ cup soy flour
2 teaspoons baking powder
Pinch sea salt
1 teaspoon vanilla extract
1 whole omega-3 egg
Grapeseed oil for griddle

In a blender, combine the banana, strawberries, and 1 teaspoon honey. Puree for 5 to 10 seconds for a chunky sauce. Set aside sauce in a small bowl.

Without washing the blender, combine the tofu, soy milk, flaxseeds, almond and soy flours, baking powder, salt, vanilla, and egg, and mix until smooth.

Preheat a griddle to 400 degrees and lightly brush with grapeseed oil. Pour approximately ¼ cup batter directly from the blender onto the griddle for each pancake. Cook pancakes until bubbles form on the surface and burst, about 4 minutes. Turn pancakes over and cook about 2 more minutes, until cooked through. Serve 3 pancakes per person with ½ cup of sauce.

Chef Tip: You can add fresh fruit to the batter and try whipped natural nut butter with a touch of honey in it as a topping.

SwiftTip: Honey, one of nature's well-studied natural remedies, is rich in antioxidants. A very small amount delivers just the right touch of sweetness.

Nutrient Notables

Pancakes, per serving (3): calories 221, carbohydrate 13g, fiber 5g, protein 14g, fat 14g, cholesterol 53mg, sodium 79mg, calcium 250mg
Sauce, per ½-cup serving, calories 66, carbohydrate 17g, fiber 3g, protein 1g, fat 0g, cholesterol 0mg, sodium 3mg, calcium 20mg

Gingered Yogurt Berry Parfait

Makes 1 serving
Prep time: 5 minutes
Cook time: None

1 cup plain yogurt (soy, cow, or goat)
¼ teaspoon dried or freshly grated ginger
1 cup fresh or frozen berries
2 tablespoons chopped nuts, any type

Combine the yogurt and ginger in a small bowl. In a parfait or other tall glass, layer the yogurt, berries, and nuts. Enjoy.

You can also top this with raw wheat germ or ground flaxseeds.

SwiftTip: The combination of yogurt, berries, and nuts is rich in fiber, minerals, and active cultures that your digestive tract will love!

Nutrient Notables

Per serving: calories 307, carbohydrate 40g, fiber 6g, protein 17g, fat 10g, cholesterol 5mg, sodium 190mg, calcium 370mg

Steel-Cut Cinnamon-Apple Oats with Flax and Nuts

Makes 4 servings
Prep time: 5 minutes
Cook time: 45 minutes

4 cups filtered water
Pinch sea salt (optional)
1 cup steel-cut oats
2 apples, skin on, cored and diced
1 teaspoon ground cinnamon
4 tablespoons flaxseeds, ground
⅓ cup chopped walnuts

Bring the water and salt (optional) to a boil and slowly stir in the oats. Boil until the oats begin to thicken, approximately 5 minutes. Reduce the heat to low, stir in the apples and cinnamon, and simmer for 30 to 40 minutes, until desired consistency. Top with ground flaxseeds and chopped walnuts.

SwiftTip: Steel-cut oats are rich in soluble or viscous fibers that help lower LDL or lousy cholesterol, and when partnered with flaxseeds, walnuts, apples, and cinnamon can boost your energy.

Nutrient Notables

Per serving: calories 282, carbohydrate 39g, fiber 8g, protein 9g, fat 11g, cholesterol 0mg, sodium 24mg, calcium 0mg

Phase II

Lunch Recipes

Dilled Egg Salad on Baby Spinach

Makes 2 servings
Prep time: 15 minutes
Cook time: 20 minutes

4 whole omega-3 eggs
2 tablespoons finely chopped scallions
2 tablespoons finely chopped fresh dill
2 tablespoons soy mayonnaise
2 teaspoons Dijon mustard
Pinch sea salt
Dash freshly ground black pepper
3 cups fresh baby spinach, trimmed and washed
1 large red apple, cut into wedges

Place the eggs in a medium saucepan and cover with cold water. Bring to a boil over medium-high heat. Remove from the heat, cover, and let stand for 15 minutes. Drain the eggs and plunge them into ice water to chill. When cold, peel and coarsely chop.

Combine the eggs, scallions, dill, mayonnaise, mustard, salt, and pepper in medium bowl and toss gently. Arrange the greens and apple wedges on a salad plate, and top with the egg salad.

SwiftTip: Dill is a flavorful and nutritious addition to protein foods such as eggs and fish, adding fiber and phytonutrients. It is soothing to the digestive tract.

Nutrient Notables

Per serving: calories 270, carbohydrate 19g, fiber 4g, protein 15g, fat 15g, cholesterol 423mg, sodium 398mg, calcium 120mg

Greek Salad Pita

Makes 2 sandwiches
Prep time: 10 minutes
Cook time: None

½ cup crumbled feta cheese
1 small pickling (Kirby) cucumber, peeled and diced
1 plum tomato, chopped
4 tablespoons chopped green bell pepper
4 tablespoons chopped red onion
4 black olives, pitted and halved
2 teaspoons extra virgin olive oil
4 teaspoons freshly squeezed lemon juice
1 tablespoon chopped fresh oregano or 1 teaspoon dried

Dressing:
½ cup plain Greek yogurt
1 tablespoon finely chopped fresh spearmint
Pinch sea salt
Dash freshly cracked black pepper

2 whole-wheat or sprouted grain pitas

Place the cheese, cucumber, tomatoes, pepper, onion, and olives in a medium bowl. Fold in the olive oil, lemon juice, and oregano. Set aside.

In a small bowl, mix together the yogurt, mint, salt, and pepper.

Slice each pita and open to form a pocket. Divide the salad between the pita pockets and drizzle with a spoonful of the yogurt dressing.

Nutrient Notables

Per pita: calories 301, carbohydrate 32g, fiber 4g, protein 13g, fat 17g, cholesterol 35mg, sodium 715mg, calcium 340mg

Wild Salmon Cakes with Asian Cucumber Cabbage Slaw

Serves: 2
Prep time: 20 minutes
Cook time: 10 minutes

One 6-ounce can wild salmon, finely chopped
½ cup red pepper, finely diced
¼ cup scallions, finely diced
½ cup celery, finely diced
¼ cup soy flour
2 tablespoons cilantro, minced
1 whole egg, omega 3-type
Pinch sea salt (optional)
Dash freshly ground black pepper (optional)
1 tablespoon sesame oil

2 cups Napa cabbage, shredded
1 seedless cucumber, thinly sliced (skin on)
½ cup Daikon radish, sliced matchstick
⅓ cup edamame beans, thawed
2 tablespoons fresh cilantro, chopped
½ teaspoon kelp powder
¼ cup unseasoned rice vinegar
2 teaspoons honey
2 teaspoons toasted sesame oil

In a large bowl, add the salmon, peppers, scallions, celery, soy flour, cilantro, egg, salt, and pepper. Mix together and form into six patties. Preheat a skillet over medium–high heat. Add sesame oil and swirl pan to coat the pan evenly. Add patties and cook for 3 to 5 minutes on each side.

Meanwhile, in a large bowl, toss together the cabbage, cucumbers, Daikon radish, edamame, cilantro, kelp powder, rice vinegar, honey, and toasted sesame oil. Divide salad between two serving plates. Place salmon cakes on top of salad. Serve.

Chef Tip: You can substitute wild salmon burgers for the homemade salmon cakes.

SwiftTip: Sea vegetables such as arame, dulse, hijiki, kelp, nori, wakame, and others are loaded with minerals, including iodine, important for thyroid function, and can add flavor to many dishes.

Nutrient Notables

Per serving: calories 279, carbohydrate 8g, fiber 3g, protein 27g,
fat 16g, cholesterol 143mg, sodium 114mg, calcium 260mg

Cucumber Slaw, per 1½-cups serving: calories 124, carbohydrate 13g, fiber
3g, protein 5g, fat 7g, cholesterol 0mg, sodium 63mg, calcium 140mg

Black Bean Confetti Salad

Makes 2 servings
Prep time: 15 minutes (plus marinating time)
Cook time: None

One 15-ounce can black beans, rinsed and drained
1 cup frozen organic corn, thawed and drained
12 cherry or grape tomatoes, halved
½ cup chopped scallions
2 cloves garlic, pressed
½ cup diced red bell pepper
¼ cup chopped cilantro
2 tablespoons extra virgin olive oil
3 tablespoons freshly squeezed lime juice
¼ teaspoon ground cumin

Mix all the ingredients in a large bowl, cover, and let marinate in the refrigerator for a few hours before serving.

Nutrient Notables

Per serving: calories 412, carbohydrate 57g, fiber 18g, protein 18g,
fat 15g, cholesterol 0mg, sodium 110mg, calcium 80mg

Quinoa Tabouleh Salad

Makes 2 servings
Prep time: 10 minutes (plus marinating time)
Cook time: 25 minutes

½ cup quinoa, thoroughly rinsed and drained
1 cup filtered water
3 tablespoons freshly squeezed lemon juice
3 tablespoons finely chopped fresh peppermint
Pinch sea salt
Dash freshly ground black pepper
3 tablespoons extra virgin olive oil
1 cup finely chopped fresh parsley
½ cup chopped scallions
2 plum tomatoes, diced
1 clove garlic, finely minced
1 cup of canned chickpeas (garbanzo beans), drained

Combine the quinoa and water in a small saucepan. Bring to a boil over medium-high heat. Cover and cook approximately 20 minutes, until the quinoa is tender and the water has been absorbed. Transfer to a medium boil and let cook slightly.

Add the lemon juice, mint, salt, pepper, olive oil, parsley, scallions, tomatoes, garlic, and chickpeas and mix all the ingredients until well coated. Cover and marinate in the refrigerator for at least 3 hours for flavors to develop.

Chef Tip: This salad is a twist on traditional tabouleh, substituting quinoa for bulgur wheat and adding chickpeas. It can be made with 1 cup of various types of cooked whole grains such as millet, amaranth, barley, kasha, or wild rice, especially if you have these leftover.

Nutrient Notables

Per serving: calories 468, carbohydrate 53g, fiber 11g, protein 14g, fat 25g, cholesterol 0mg, sodium 88mg, calcium 150mg

Wasabi Salmon Salad

Makes 2 servings
Prep time: 10 minutes
Cook time: None

One 6-ounce can wild salmon, drained
2 tablespoons minced scallions
1 tablespoon red pepper, diced
1 tablespoon diced celery
1 teaspoon grated fresh ginger
½ cup plain Greek yogurt
¼ teaspoon wasabi powder, or to taste
2 cups chopped Napa cabbage or bok choy

In a medium bowl, mix together the salmon, scallions, celery, ginger, red pepper, yogurt, and wasabi. Serve on chopped Napa cabbage or bok choy.

Nutrient Notables

Per serving: calories 225, carbohydrate 17g, fiber 3g, protein 22g,
fat 8g, cholesterol 37mg, sodium 164mg, calcium 370mg

Curried Turkey Salad on Watercress

Makes 2 servings
Prep time: 15 minutes
Cook time: None

6 to 8 ounces cooked turkey breast, julienned
½ cup celery, chopped
¼ cup red onion, chopped
½ cup red seedless grapes
½ cup walnuts, chopped
½ cup plain nonfat yogurt
1 teaspoon ground curry powder
2 bunches watercress, washed and trimmed

In a medium bowl, mix the turkey, celery, red onion, walnuts, grapes, yogurt, and curry powder together. Serve on a bed of watercress.

SwiftTip: Curry powder is a blend of many different anti-inflammatory herbs, spices, and seeds, such as cardamom, chilies, cinnamon, cloves, coriander, fennel, fenugreek, mace, nutmeg, pepper, saffron, turmeric, and others.

Nutrient Notables

Per serving: calories 409, carbohydrate 20g, fiber 4g, protein 40g, fat 20g, cholesterol 84mg, sodium 143mg, calcium 250mg

Phase II

Dinner Recipes

Pan-Seared Scallops with Peach Kiwi Salsa, Baby Bok Choy, and Steamed Kasha

Makes 4 servings
Prep time: 20 minutes
Cook time: 10 minutes

1 cup buckwheat groats (kasha)
3 cups low-sodium organic vegetable broth
½ cup minced red onion
1 cup frozen peaches, thawed and diced small
3 kiwis, peeled and diced
1 teaspoon honey
1 teaspoon freshly squeezed lime juice
¼ cup chopped fresh cilantro
Pinch sea salt
1 tablespoon sesame oil
1 pound fresh sea scallops, patted dry
4 heads baby bok choy

Combine the kasha and broth in a medium saucepan. Bring to a boil on medium-high heat, cover, reduce the heat to low, and simmer for 15 to 20 minutes. Set aside to keep warm.

In a medium bowl, combine the onion, peaches, kiwi, honey, lime juice, cilantro, and salt. Mix well and set aside.

Heat a medium skillet over medium heat until hot. Add the oil and swirl the pan to evenly distribute the oil. Add the scallops one at a time in a single layer. Let scallops cook undisturbed for 2 to 3 minutes, until brown on the bottom. Carefully turn over the scallops and cook on the other side for 3 minutes more. Steam the baby bok choy. Divide the bok choy among 4 serving plates. Place the cooked kasha and scallops on top of the bok choy, and finish with ¼ cup of salsa.

SwiftTip: Kiwis are rich in fiber, potassium, and vitamin C and are a nutrient dense addition to salads, salsas, and chutneys.

Nutrient Notables

Scallops and salsa, per serving: calories 272, carbohydrate 27g, fiber 3g, protein 29g, fat 5g, cholesterol 64mg, sodium 324mg, calcium 160mg
Buckwheat groats, per ½-cup serving: calories 77, carbohydrate 17g, fiber 2g, protein 3g, fat 1/0g, cholesterol 0mg, sodium 3mg, calcium 10mg
Baby bok choy, per 1-cup serving: calories 28, carbohydrate 5g, fiber 2g, protein 2g, fat 0g, cholesterol 0mg, sodium 30mg, calcium 210mg

Blackstrap Vegetarian Chili with Baby Greens Salad and Citrus Vinaigrette

Makes 4 servings
Prep time: 20 minutes
Cook time: 30 minutes

One 15-ounce can kidney beans
One 15-ounce can pinto beans
Two 14.5-ounce cans diced tomatoes with juice
½ cup chopped celery
1 cup diced yellow onion
2 tablespoons chopped fresh garlic
1 cup peeled, diced delicata squash
½ cup diced red pepper
2 cups low-sodium organic vegetable broth
2 tablespoons blackstrap molasses
3 tablespoons chili powder
1 tablespoon ground cumin
¼ cup extra virgin olive oil

2 tablespoons freshly squeezed lemon juice
2 tablespoons balsamic vinegar
Pinch sea salt (optional)
Freshly ground pepper
Chopped fresh basil or other herb to taste
2 teaspoons Dijon mustard
8 cups mixed baby greens

Combine the beans, tomatoes, celery, onion, garlic, squash, red peppers, broth, molasses, chili powder, cumin, and salt in a large saucepan. Cover and simmer over medium-low heat for 20 minutes, until the vegetables are cooked through.

Whisk together the oil, lemon juice, and basil. Pour over the greens in a large bowl and toss to coat. Serve the chili with the green salad.

Chef Tip: Make a double batch of this easy chili, freeze in individual portions, and you will have it on hand for those extra-busy days.

SwiftTip: Blackstrap molasses, one of grandmother's favorite ingredients and a natural remedy, adds valuable minerals such as calcium, potassium, and iron to your diet.

Nutrient Notables

Chili, per 1-cup serving: calories 381, carbohydrate 77g, fiber 24g,
protein 19g, fat 3g, cholesterol 0mg, sodium 1060mg, calcium 310mg
Salad, per 2-cup serving: calories 50, carbohydrate 10g, fiber 3g,
protein 3g, fat 0g, cholesterol 0mg, sodium 69mg, calcium 70mg
Citrus vinaigrette, per 1-tablespoon serving: calories 63, carbohydrate 1g,
fiber 0g, protein 0g, fat 7g, cholesterol 0mg, sodium 14mg, calcium 0mg

Chef Tip: You can reduce the sodium by rinsing canned beans.

Raspberry-Pistachio Crusted Chicken with Steamed Kale and Wild Rice

Makes 4 servings
Prep time: 15 minutes
Cook time: 55 minutes

½ cup wild rice
1½ cups filtered water

1 cup fresh or (unsweetened) frozen raspberries
1 tablespoon Dijon mustard
2 tablespoons freshly squeezed lemon juice

4 skinless, boneless chicken breast halves, about 1 pound
½ cup whole-grain bread crumbs
2 tablespoons coarsely ground pistachio nuts
2 tablespoons minced fresh parsley
Dash freshly ground white pepper
Pinch sea salt
1 head kale (about 1 pound), leaves trimmed and washed
2 teaspoons olive oil

Put the wild rice, water, and a pinch of salt in a medium saucepan and bring to a boil. Reduce the heat and simmer, covered for 55 minutes.

While the rice is cooking, combine the raspberries, mustard, and lemon juice in a small food processor or blender. Process until smooth. Transfer to a shallow pan, cover, and set aside.

Place the chicken breasts between two sheets of wax paper and pound with a meat mallet to ½-inch thick. In another shallow pan, combine the bread crumbs, pistachios, parsley, pepper, and a pinch of salt. Dip each chicken breast in the sauce to coat both sides. Roll the chicken in the breadcrumb mixture to cover completely.

Steam the kale in a large covered sauce pan until tender. Set aside to keep warm.

Heat the olive oil in a large skillet and sauté the chicken breasts over medium heat until the chicken is cooked through and the crust is slightly browned, 5 minutes on each side. Serve with the steamed kale and wild rice.

Nutrient Notables

Raspberry-pistachio crusted chicken per serving: calories 246, carbohydrate 14g, fiber 4g, protein 30g, fat 8g, cholesterol 65mg, sodium 193mg, calcium 50mg

Steamed kale, per 1-cup serving: calories 35, carbohydrate 7g, fiber 2g, protein 2g, fat 0g, cholesterol 0mg, sodium 30mg, calcium 90mg

Wild rice, per half-cup serving: calories 83, carbohydrate 17g, fiber 3g, protein 7g, fat 1g, cholesterol 0mg, sodium 112mg, calcium 0mg

Mustard-Crusted Lamb Chops with Garlicky Broccoli Rabe

Makes 2 servings
Prep time: 20 minutes
Cooking time: 10 minutes

1 tablespoon Dijon mustard
1 tablespoon finely chopped fresh rosemary
1 tablespoon minced shallots
2 tablespoons whole-grain bread crumbs
2 tablespoons minced garlic
2 tablespoons extra virgin olive oil
Six 2- to 4-ounce lean loin lamb chops
3 cups chopped broccoli rabe (about ¾ pounds, trimmed)

Preheat the oven to 475 degrees.

In a small bowl, make a paste with the mustard, rosemary, shallots, bread crumbs, 1 tablespoon of the garlic, and 1 tablespoon of the olive oil. Rub the paste on both sides of the lamb chops, spreading evenly to cover the meat. Bake chops for 5 minutes on each side or until desired degree of doneness. Place a piece of parchment paper on a baking sheet. Lay the chops on the paper about 2 inches apart.

Meanwhile, heat a large sauté pan over medium heat. Add the remaining olive oil and garlic. Add the broccoli rabe and toss in the pan to lightly coat with the oil and garlic. Cook for 3 to 5 minutes until wilted and bright green. Serve with the lamb chops.

SwiftTip: Broccoli rabe or rapini is a great, fiber-rich green that has a mustardy bite. It helps with liver detoxification.

Nutrient Notables

Lamb chops (1 chop), per serving: calories 266, carbohydrate 8g, fiber 1g, protein 26g, fat 14g, cholesterol 72mg, sodium 210mg, calcium 50mg
Broccoli rabe, per 1-cup serving: calories 135, carbohydrate 7g, fiber 6g, protein 9g, fat 8g, cholesterol 0mg, sodium 129mg, calcium 270mg

Cajun-Spiked Halibut with Peppers, Mustard Greens, and Baked Sweet Potatoes

Makes 2 servings
Prep time: 5 minutes
Cook time: 45 minutes

2 small sweet potatoes, scrubbed and pierced
1 teaspoon paprika
1 tablespoon minced fresh oregano or 1 teaspoon dried
1 tablespoon minced fresh thyme or 1 teaspoon dried
½ teaspoon cayenne pepper
Pinch sea salt

Two 4- to 6-ounce Alaskan halibut steaks
1 tablespoon olive oil
1 red bell pepper, seeded and thinly sliced
1 green bell pepper, seeded and thinly sliced
1 bunch mustard greens (about ½ pound), trimmed, rinsed, and chopped

Preheat the oven to 400 degrees. Bake the potatoes until soft, about 45 minutes.

Combine the paprika, oregano, thyme, cayenne, and salt in a small bowl. Place the halibut on a plate and sprinkle with the seasoning to coat the fish well. Heat 2 teaspoons of the olive oil in a skillet, add the peppers and mustard greens, and sauté on medium heat for 5 to 7 minutes. Add the remaining teaspoon of olive oil and the halibut. Sauté over medium heat for 3 to 4 minutes on each side until browned and cooked through. Serve the fish and vegetables with a baked sweet potato.

Nutrient Notables

Fish and vegetables, per serving: calories 388, carbohydrate 10g, fiber 6g, protein 26g, fat 28g, cholesterol 68mg, sodium 145mg, calcium 120mg
Sweet potato, each: calories 100, carbohydrate 24g, fiber 4g, protein 2g, fat 0g, cholesterol 0mg, sodium 41mg, calcium 40mg

Chicken-Vegetable Fajitas in Sprouted Grain Tortillas

Makes 4 servings
Prep time: 20 minutes (plus marinating time)
Cook time: 15 minutes

2 tablespoons freshly squeezed lime juice
½ teaspoon chili powder
½ teaspoon ground cumin
1 clove garlic, minced
1 pound skinless, boneless chicken breasts, sliced into ½-inch strips,
    or 1 pound chicken tenders
2 tablespoons olive oil
1 medium green bell pepper, thinly sliced
1 medium red bell pepper, thinly sliced
1 medium yellow bell pepper, thinly sliced
1 large yellow onion, thinly sliced
Four 10-inch sprouted or whole-grain tortillas

Optional garnishes
Avocado
Chopped green chiles
Low-fat sour cream
Salsa
Shredded cheese

Combine the lime juice, chili powder, cumin, and garlic in a medium bowl and mix well. Add the chicken and marinate, covered, in the refrigerator for at least 3 hours.

Heat the oil in a large skillet over medium-high heat and cook the chicken and vegetables, stirring occasionally, until the chicken is cooked through and the vegetables are crisp-tender, about 15 minutes. Divide the mixture among the tortillas.

SwiftTip: You can substitute hickory-smoked tofu or tempeh in place of the chicken in this recipe for a delicious vegetarian meal.

Nutrient Notables

Per serving: calories 404, carbohydrate 46g, fiber 7g, protein 35g,
fat 10g, cholesterol 68mg, sodium 422mg, calcium 40mg

Dr. Hyman's Chinese Eggs and Seasoned Greens

Makes 4 servings
Prep time: 5 minutes
Cook time: 45 minutes

1 cup brown rice, rinsed
2 cups filtered water
12 garlic cloves
6 whole omega-3 eggs
3 tablespoons olive oil
One 14.5-ounce can chopped plum tomatoes, low sodium
 with juice
1 teaspoon toasted sesame oil
2 teaspoons reduced-sodium, wheat-free tamari
1 teaspoon Worcestershire sauce
12 cups spinach, rinsed but not dried

Combine the brown rice and water in a small saucepan. Bring to a boil, lower the heat, and cook approximately 45 minutes until the rice is tender and the water has been absorbed.

While the rice is cooking, chop the garlic coarsely. Beat the eggs with a whisk. Heat the olive oil in a large nonstick sauté pan or wok at medium heat. Add the garlic and cook for 1 minute. Add the eggs and let them cook undisturbed until no longer liquid, about 3 to 5 minutes, then flip over and cook on the other side. When cooked through, 2 to 3 minutes, cut into pieces about 1 to 2 inches square. Add the chopped tomatoes and juice to the eggs. Add the sesame oil, tamari, and Worcestershire sauce. Simmer for 10 minutes.

While the eggs are simmering, steam the spinach in a large covered saucepan. Serve the eggs and sauce over the brown rice with some steamed spinach.

SwiftTip: Egg yolks, once shunned for their cholesterol content, are rich in many nutrients including choline, a phospholipid that is a key component of cells and important for a healthy nervous system.

Nutrient Notables

Eggs and sauce per serving: calories 250, carbohydrate 10g, fiber 1g, protein 11g, fat 19g, cholesterol 317mg, sodium 221mg, calcium 100mg

Steamed spinach, per 2-cup serving: calories 82, carbohydrate 14g, fiber 8g, protein 5g, fat 0g, cholesterol 0mg, sodium 126mg, calcium 240mg

Brown rice, per ½-cup serving: calories 108, carbohydrate 45g, fiber 4g, protein 5g, fat 2g, cholesterol 0mg, sodium 10mg, calcium 20mg

Phase II

Snacks

Lemony Hummus

Makes about 1¼ cups hummus
Prep time: 15 minutes
Cook time: none

One 15-ounce can chickpeas, well drained
¼ cup tahini
1 garlic clove, finely chopped
½ cup lemon juice, freshly squeezed
1 tablespoon extra virgin olive oil
2 tablespoons water
Sea salt and freshly ground black pepper (optional)

In a food processor, place the chickpeas, tahini, garlic, lemon juice, olive oil, and water and puree until thick and creamy. Season to taste with salt and pepper. Adjust consistency with additional lemon juice if desired.

Serve on raw vegetables (baby carrots, jicama, red pepper, celery, Daikon radish).

Nutrient Notables

Hummus, per 2-tablespoon serving: calories 93, carbohydrate 10g, fiber 3g, protein 4g, fat 5g, cholesterol 0mg, sodium 98mg, calcium 40mg

Raw vegetables, per 1-cup serving: calories 40, carbohydrate 9g, fiber 3g, protein 2g, fat 0g, cholesterol 0mg, sodium 20mg, calcium 50mg

Yogurt with Fresh Fruit and Shredded Coconut and Wheat Germ

Makes 1 serving
Prep time: 5 minutes
Cook time: none

One 8-ounce carton plain yogurt
1 cup fruit in season: kiwi, strawberries, mango, diced
2 teaspoons grated coconut
1 teaspoon raw wheat germ

Mix fruit in yogurt and top with grated coconut and wheat germ.

Nutrient Notables

Per 2-cup serving: calories 275, carbohydrate 43g, fiber 5g, protein 17g,
fat 5g, cholesterol 5mg, sodium 194mg, calcium 320mg

Antipasto

Makes 1 serving
Prep time: 5 minutes
Cook time: none

8 to 10 olives, black or green
½ cup roasted red peppers, drained
2 artichoke hearts, drained
4 hearts of palm, drained

Arrange vegetables on a plate and serve.

Nutrient Notables

Per serving: calories 131, carbohydrate 20g, fiber 10g, protein 7g,
fat 5g, cholesterol 0mg, sodium 520mg, calcium 170mg

Almond Butter with Flax Crackers

Makes 1 serving
Prep time: 5 minutes
Cook time: None

Almond butter is a delicious spread that is made from ground almonds and can be substituted for peanut butter. It can be used as an ingredient in recipes or as a spread on whole-grain crackers or fresh fruit such as apple slices.

Nutrient Notables

Almond butter, per 1-tablespoon serving: calories 100, carbohydrate 3g, fiber 1g, protein 2g, fat 9g, cholesterol 0mg, sodium 2mg, calcium 40mg
Flax crackers (Mary's Gone Crackers), per 15-cracker serving: calories 120, carbohydrate 20g, fiber 3g, protein 3g, fat 3.5g, cholesterol 0mg, sodium 168mg, calcium 40mg

UltraMetabolism Road Mix

Makes thirteen ½-cup servings
Prep time: 5 minutes
Cook time: None

½ cup dried wild blueberries
1 cup cocoa nibs
1 cup raw almonds, whole
1 cup raw cashews, whole
1 cup raw walnuts, whole
1 cup hulled raw pumpkin seeds
1 cup hulled raw sunflower seeds

In a medium bowl, mix all the ingredients. Store in a covered jar and keep in a cool, dark place.

Chef Tip: Cocoa nibs are roasted cocoa beans separated from their husks and broken into small bits. The nibs can be used in recipes or as a stand-alone snack when nothing but chocolate will satisfy your taste buds.

Nutrient Notables

Per ½-cup serving: calories 300, carbohydrate 23g, fiber 10g, protein 13g, fat 24g, cholesterol 0mg, sodium 54mg, calcium 70mg

Brazil Nut Bars

Makes 16 pieces
Prep time: 15 minutes
Cook time: 3 minutes

Grapeseed oil
1½ cups whole Brazil nuts
½ cup hulled raw pumpkin seeds
½ cup sliced almonds
½ cup hulled raw sunflower seeds
½ cup ground flaxseeds
⅓ cup dried organic cranberries
1 teaspoon ground cinnamon
1½ cups high-fiber, low-sugar sprouted grain cereal*
¾ cup honey
1 cup natural cashew butter

Lightly coat a 9x13-inch baking dish with grapeseed oil. Set aside.

In the bowl of a food processor, pulse the Brazil nuts until the nuts are ground into a fine powder. Transfer to a large bowl. Add the pumpkin seeds, almonds, sunflower seeds, flaxseeds, cranberries, cinnamon, and sprouted grain cereal. In a large saucepan on the stove, heat the honey and cashew butter on medium high until very hot and bubbling, about 3 minutes. Pour over the mixture in the bowl and mix well with a wooden spoon. Immediately, press mixture firmly into the baking dish (wear rubber gloves if needed). Let the mixture cool in the refrigerator. Cut into 16 pieces. Wrap each bar individually in wax paper and store in the freezer.

Chef Tip: These bars make an excellent snack or a quick, on-the-run breakfast with a piece of fruit.

Nutrient Notables

Per serving: calories 377, carbohydrates 30g, fiber 5g, protein 10g, fat 27g, cholesterol 0mg, sodium 25mg, calcium 170mg

---

* Try using Food for Life Ezekiel 4:9 Original sprouted grain cereal.

Phase II

Desserts

Simple Dessert Ideas:

> Fresh fruit dipped in all-natural dark chocolate sauce
> Baked Fruit (below)
> Plain yogurt with fruit and shredded coconut
> Brazil Nut Bar (p. 305)

Baked Fruit

Makes 3 servings
Prep time: 10 minutes
Cook time: 15 minutes

Fresh fruit in season, such as apples, pears, peaches, plums, and apricots '
1 tablespoon balsamic vinegar
¼ teaspoon ground cardamom

Preheat the oven to 375 degrees.

Cut the fruit into 1- to 2-inch cubes. You'll need 4 cups of cubes. Place in a shallow baking dish. Drizzle with the balsamic vinegar and sprinkle with the cardamom. Bake for 15 minutes or until fruit is tender.

Nutrient Notables

Per 1-cup serving: calories 150, carbohydrate 30g, fiber 4g, protein 2g, fat 0g, cholesterol 0g, sodium 0g, calcium 10mg

Serious Hot Chocolate*

Makes 1 serving
Prep time: 5 minutes (plus 20 minute rest)
Cook time: 10 minutes

1 cup unsweetened soy milk or filtered water
4 to 5 tablespoons chopped dark chocolate or Chocolate Springs Mix
Ground cinnamon or shaved chocolate, for garnish

---

* Recipe adapted from chef Joshua Needleman, creator of Chocolate Springs Café in Lenox, Massachusetts (see Resources)

Heat the liquid and pour approximately ¼ cup of it over the chocolate in a small bowl. Whisk to a smooth paste. Pour the chocolate paste back into the hot liquid and whisk while bringing just to a boil. Remove from the heat and let it rest approximately 20 minutes, then reheat it. Serve with cinnamon or chocolate shavings.

The finished drink will last about 5 days in the refrigerator.

Nutrient Notables

Per 1-cup serving: calories 430, carbohydrate 45g, fiber 7g, protein 14g, fat 24g, cholesterol 0mg, sodium 138mg

Fruit-Filled *Kanten* (Japanese Jell-o) with Almond Mousse

Makes 12 servings
Prep time: 15 minutes (plus jelling time)
Cook time: 2 minutes

4 cups 100% pomegranate juice
1 cup filtered water
5 tablespoons powdered agar-agar
1 teaspoon vanilla extract
1 pint fresh blueberries
½ mango, peeled and sliced
3 kiwis, peeled and sliced
1 tablespoon natural almond butter
½ cup unsweetened soy milk

Place the pomegranate juice and water into a saucepan and bring to a boil. Lower the heat to medium and add the agar-agar and the vanilla, stirring constantly until dissolved. Line a medium dessert bowl with the fruit. When the agar-agar mixture has cooled, pour half of it into the bowl with the fruit and pour remaining into a separate bowl.

Chill until set. Remove the set mixture with the fruit and place into a blender. Add the almond butter and soy milk and blend until creamy.

To serve, slice the mold and top with a dollop of mousse.

Nutrient Notables

Per serving: calories 90, carbohydrate 18g, fiber 1g, protein 2g, fat 2g, cholesterol 0mg, sodium 12mg, calcium 50mg

Almond Macaroons

Makes about 18 pieces
Prep time: 10 minutes
Cook time: 20 to 25 minutes

Walnut or grapeseed oil (to grease cookie sheet)
2⅔ cups unsweetened shredded coconut
1 cup sliced raw almonds
¼ cup agave nectar
4 large omega-3 egg whites

Preheat oven to 325 degrees. Lightly oil 2 large cookie sheets.

In a large bowl, mix the coconut, almonds, and agave nectar until combined. Stir in the egg whites and blend well. Drop mixture by heaping tablespoonfuls, about 2 inches apart on cookie sheets.

Bake cookies until golden, 20 to 25 minutes. Remove cookies to wire racks to cool. Store in a tightly covered container.

Nutrient Notables

Per serving: calories 122, carbohydrate 8g, fiber 2g, protein 2g,
fat 10g, cholesterol 0mg, sodium 16mg, calcium 20mg

Best Brownie Ever

Makes 12 servings
Prep time: 10 minutes
Cook time: 20 minutes

1 cup raw pecans
6 tablespoons walnut oil plus extra for the baking pan
½ cup agave nectar
2 whole omega-3 eggs
½ cup cocoa powder
¼ cup arrowroot

Preheat oven to 350 degrees. Oil an 8x8x2-inch baking pan.

In a food processor, grind the pecans to the consistency of meal. Transfer to a medium bowl and add the walnut oil, agave nectar, eggs, cocoa, and arrow-root. Stir to blend. Pour into an oiled 8x8x2-inch baking dish.

Bake for 20 minutes or until a toothpick comes out clean. Let cool, then cut into 12 bars.

Nutrient Notables

Per serving: calories 197, carbohydrate 17g, fiber 2g, protein 3g, fat 15g, cholesterol 35mg, sodium 13mg, calcium 20 mg

Hint-of-Orange Carrot Cake

Makes 12 servings
Prep time: 20 minutes
Cook time: 50 minutes

Walnut oil
6 omega-3 eggs, separated
½ cup agave nectar
1½ cups pureed carrots (about 6 to 8 medium carrots)
2 tablespoons grated orange zest
1 tablespoon frozen orange juice concentrate
1 teaspoon ground ginger
3 cups almond meal

Preheat oven to 325 degrees. Oil the bottom of a 9-inch springform pan.

Beat the egg yolks and agave nectar together. Stir in the carrot puree, orange zest, orange juice, ginger, and almond meal. Beat the egg whites until stiff peaks form and fold gently into the carrot mixture. Spoon into the prepared springform pan.

Bake for approximately 50 minutes, until a knife inserted into the center of the cake comes out clean. Cool in the pan for about 15 minutes and then turn onto a wire rack to cool completely.

Nutrient Notables

Per serving: calories 230, carbohydrate 20g, fiber 4g, protein 9g, fat 15g, cholesterol 106mg, sodium 42mg, calcium 60mg

Macadamia Muffins

Makes 16 muffins
Prep time: 10 minutes
Cook time: 15 minutes

1 cup natural macadamia nut butter
1 cup sliced raw nuts (any variety)
1 cup coconut milk
2 cups unsweetened shredded coconut
3 whole omega-3 eggs
½ teaspoon ground allspice
⅓ cup chopped dried organic fruit such as raisins, blueberries, cranberries, or mango (optional)

Preheat the oven to 400 degrees.

Combine the macadamia butter, nuts, coconut milk, coconut, eggs, allspice, and fruit (if using) in a medium bowl and stir until well blended. Spoon batter into 16 unbleached muffin liners. Bake for 15 minutes.

Nutrient Notables (without optional dried fruit)

Per muffin: calories 225, carbohydrate 5g, fiber 3g, protein 3g, fat 23g, cholesterol 35mg, sodium 17mg, calcium 30 mg

Nutrient Notables (with about ⅓ cup dried fruit)

Per muffin: calories 235, carbohydrate 7g, fiber 3g, protein 4g, fat 23g, cholesterol 35mg, sodium 17mg, calcium 30mg

Phase II Shopping List

Vegetables

Arugula, ½ pound
Baby bok choy, 4 heads, ¼ pound
Baby mixed greens
Baby spinach, ½ pound
Broccoli rabe, 2 bunches
Carrots, 1 bag baby carrots
Celery, 1 bunch

Cucumbers, 11
Daikon radish, 1
Delicata squash, 1
Garlic, 2 bulbs
Ginger, fresh, 3-inch piece
Herbs (bunch): Parsley 2, Dill 1,
     Oregano 1, Peppermint 1,

Rosemary 1, Thyme 1,
  Cilantro 1, Spearmint 1
Kale, 1 head
Mustard greens, 1 bunch
Napa cabbage, 2 heads
Olives, 8 green and 4 black
Onions, 2 yellow and 2 red
  onions
Peppers, green bell, 3

Peppers, red bell, 4
Peppers, yellow bell, 1
Scallions, 2 bunches
Shallots, 3
Spinach, 1½ pounds
Sweet potatoes, 2
Tomatoes, grape 1 package and
  3 plum tomatoes
Watercress, 2 bunches

Fruit

Apples, 4 any variety
Avocado, 1
Bananas, 2
Berries (any type), 3 pints
Blueberries, 1 pint
Grapes, red seedless, 1 bunch
Kiwi, 3

Lemons, 5
Limes, 2
Oranges, 2
Pears, 3
Raspberries, ½ pint
Strawberries, 1 pint

Nuts and Seeds

Almonds, 1 pound
Brazil nuts, 1 pound
Cashews, ¼ pound
Pistachios, ¼ pound

Pumpkin seeds, ½ pound
Sunflower seeds, ½ pound
Walnuts, ½ pound

Meat, Fish, and Poultry

Chicken breasts, skinless and
  boneless, 1 pound, or chicken
  tenders
Turkey breast, 6 to 8 ounces
Lamb loin chops, two 4-ounce
  chops

Alaskan halibut, two 4- to
  6-ounce steaks
Shrimp, 8 ounces large size
Sea scallops, 1 pound

Eggs and Dairy

Omega-3 organic eggs, 2 dozen

Feta cheese, 4 ounces

Organic Soy Products

Plain unsweetened soy milk,
1 quart

Plain unsweetened soy yogurt,
one 8-ounce carton

Silken tofu, 1 small
package

Dairy

Greek plain nonfat yogurt, two
8-ounce cartons

Plain nonfat yogurt, two
8-ounce cartons

Frozen Foods

One 10-ounce package frozen
mixed berries

One 10-ounce package frozen
corn

One 10-ounce package frozen
edamame beans

One 10-ounce package frozen
peaches

Whole Grains and Whole-Grain Products

Steel-cut oats, 1 cup (1 box or
canister)

Whole rye or flax bread, 3 slices
(dense German type)

Sprouted grain cereal, 2¼ cups
(1 box)

Pitas, whole grain or sprouted
grain, 2

Sprouted grain tortillas, 4

Whole or sprouted grain bread
crumbs, ¾ cup

Wheat germ, raw, 1 box

Pantry

Artichoke hearts, 1 jar

Beans, canned: Chickpeas
(garbanzo) 1 can, black 1 can,
kidney 1 can, pinto 1 can

Coconut, grated, 2 teaspoons

Dark chocolate, 1½ pounds

Dried wild blueberries, ½ cup

Dried cranberries, ⅓ cup

Hearts of palm, 1 jar

Roasted red peppers, 1 jar

Tomatoes, diced, two
14.5-ounce cans

Tomatoes, plum, one 16-ounce
can

Wild salmon, two 6-ounce cans

Staples

Almond flour, 1-pound
  package
Baking powder, 1 can
Blackstrap molasses, 1 jar
Cocoa nibs, 1 cup
Coconut milk, light, 1 can
Grapeseed oil, 1 small bottle
Soy flour, 1 small package
Soy mayonnaise, 1 small jar

Tamari, low sodium, 1 bottle
Vanilla extract, 1 small bottle
Vegetable broth, low sodium,
  organic, 1 pint
Worcestershire sauce
  (vegetarian), 1 small bottle

Herbs, Spices, and Seasonings

Cayenne pepper
Kelp powder
Oregano, dried

Paprika
Thyme, dried
Wasabi powder

CONCLUSION

Parting Thoughts: Reflections on the Past and the Future

The last twenty years have been a journey of discovery for me. I have learned more from my patients and their struggles by asking questions and listening carefully than from any book.

Most of them could tell me exactly what their problems were, even if they didn't know exactly why things were not working, or how to fix them. Their patience, persistence, and belief in themselves, and their desire for health and feeling well, inspired me to find the answers to their problems. I asked more and more questions until I found their answers.

Because of them, I read, studied, asked experts, and pored over the research. The results of my inquiry allowed me to make the connections between the scientific research and their suffering, and to find some important solutions. It also allowed me to write this book.

This book is for all of them, and for the children who truly are victims whose fate will not be fully told for a few decades. Urgent change is needed to reverse our declining life-expectancy and the specter of an economy bankrupt from the health costs of obesity because we would not take the risks now to make difficult, politically costly changes to our food policy and food environment.

The story of obesity is still being told, and there are many exciting discoveries about health and metabolism ahead of us. Nutrigenomics is in its infancy. However, the study of how diet influences genes, and how we can use that information to stop our health-care crisis and improve health for everyone, is complete enough now to dramatically change the obesity and health-care crisis we face. We do not have to leave our children with the legacy of sicker, shorter lives.

When I think about the vast crisis we are facing now, a crisis that goes to the heart of our health, civilization, economy, and our "pursuit of happiness," I am reminded of something my friend Michael Shane told me recently. It is something I truly believe about our obesity and health-care crisis.

"There is enough information in the world to solve any problem."

You'll find a vast array of organic food products, home-care, health-care, kitchenware, and other valuable resources from these sites:

Organic Essentials
The Organic Food Pages
www.theorganicpages.com

Oraganics
www.oraganic.com

EfoodPantry.com
www.efoodpantry.com

Organic Provisions
www.orgfood.com

Organic Planet
www.organic-planet.com

Sun Organic Farm
www.sunorganicfarm.com

Green for Good
www.greenforgood.com

Produce
Diamond Organics
www.diamondorganics.com
  Mail order high-quality organic produce and raw foods

Earthbound Farms
www.earthboundfarm.com
  Fresh packaged organic produce

Small Planet Foods
www.cfarm.com
  Home site of Cascadian Farms and Muir Glen, organic frozen and canned vegetables and fruits

Maine Coast Sea Vegetables
www.seaveg.com
  Variety of sea vegetables, including some organically certified types

Meat, Poultry, Eggs, and Dairy
Eat Wild
www.eatwild.com
  Information and ordering site for grass-fed meat and dairy products

Organic Valley
www.organicvalley.com
  Organic meats, dairy, eggs, and produce from more than 600 member-owned organic farms

Peaceful Pastures
www.peacefulpastures.com
  Grass-fed and grass-finished meat, poultry, and dairy products

Applegate Farms
www.applegatefarms.com
  Packaged poultry, meats, and deli products

Stonyfield Farm
www.stonyfield.com
  Certified organic dairy products
  and soy yogurt

Horizon Organic
www.horizonorganic.com
  Variety of certified organic dairy
  products including cheeses

Fish
Vital Choice Seafood
www.vitalchoice.com
  Selection of wild fresh, frozen, and
  canned salmon

Ecofish, Inc.
www.ecofish.com
  Environmentally responsible
  seafood products and
  information

Crown Prince, Inc.
www.crownprince.com
  Wild-caught, sustainably
  harvested specialty canned
  seafood

Sea Bear Smokehouse
www.seabear.com
  Wild salmon jerky for a convenient snack

Nuts, Seeds, and Oils
Barlean's Organic Oils
www.barleans.com
  Organic oils and ground flaxseed

Omega Nutrition
www.omeganutrition.com
  Variety of organic oils, flax, and
  hempseed products

Spectrum Naturals
www.spectrumorganic.com
  Extensive line of high-quality oils,
  vinegars, flax products, and culinary resources

Maranatha
www.nspiredfoods.com
  Organic nut and seed butters

Once Again Nut Butter
www.onceagainnutbutter.com
  Organic nut and seed butters

Beans and Legumes
Eden Foods
www.edenfoods.com
  Complete line of organic dried
  and canned beans

Westbrae Natural
www.westbrae.com
  Full variety of organic beans
  and vegetarian products
  (soups, condiments,
  pastas, etc.)

ShariAnn's Organic
www.shariannsorganic.com
  Organic beans, refried beans,
  soups, and more

Grains
Arrowhead Mills
www.arrowheadmills.com
  Organic grains including many
  gluten-free choices

Lundberg Family Farms
www.lundberg.com
  Organic grains and gluten-free
  items such as wild rice

Alvarado St. Bakery
www.alvaradostreetbakery.com
  Organic sprouted grain products

French Meadow Bakery
www.frenchmeadow.com
  Organic sprouted grain products

Carl Brandt, Inc
www.carlbrandt.com
  Dense German-style breads
  including flaxseed and others

Hodgson Mill, Inc.
www.hodgsonmill.com
  Complete line of whole grains
  including many gluten-free grains

Shiloh Farms Bakery
www.shilohfarms.net
  Organic whole grains, sprouted
  grains, and gluten-free items

Tinkyáda
www.tinkyada.com
  Wheat- and gluten-free pastas

Spices, Seasonings, Sauces, Soups, and Such

Spice Hunter
www.spicehunter.com
  Complete line of organic spices

Frontier Natural Products Co-op
www.frontiernaturalbrands.com
  Extensive line of organic spices,
  seasonings, baking flavors and extracts, dried foods, teas, and culinary gadgets

Rapunzel Pure Organics
www.rapunzel.com
  Great selection of seasonings such
  as Herbamare made with sea salt
  and organic herbs

Seeds of Change
www.seedsofchange.com
  Organic tomato sauces, salsas, and
  more

Edward and Sons Trading
Company, Inc.
www.edwardandsons.com
  Extensive line of vegetarian organic
  food products including miso,
  sauces, brown rice crackers, etc.

Pacific Natural Foods
www.pacificfoods.com
  Organic and gluten-free broths,
  soups, ready-to-eat meals, and
  beverages

Imagine Foods
www.imaginefoods.com
  Certified organic gluten-free
  stocks and soups

Flavorganics
www.flavorganics.com
  Full product line of certified organic pure flavor extracts

Beverages

Nondairy, Gluten-Free Beverages

Westbrae Westsoy (unsweetened
soy milk)
www.westbrae.com

Imagine Foods (Soy Dream)
www.imaginefoods.com

WhiteWave (Silk soy beverage)
www.whitewave.com

Pacific Natural Foods (Select soy,
almond, hazelnut, and rice
beverages)
www.pacificfoods.com

Whole Soy Co. (unsweetened soy yogurt)
www.wholesoyco.com

Lundberg Family Farms (Drink Rice beverage)
www.lundberg.com

Organic Herbal Teas

Choice Organic Teas
www.choiceorganicteas.com

Yogi Teas
www.yogitea.com

Republic of Tea
www.republicoftea.com

Numi Tea
www.numitea.com

Organic Spirits

Frey Vineyards
www.freywine.com

Fetzer Vineyards
www.fetzer.com

Chocolate

Chocolate Springs
www.chocolatesprings.com
    A wide range of high-quality chocolate items including chocolate nibs

Dagoba Organic Chocolate
www.dagobachocolate.com
    Premium organic chocolate products including raw chocolate nibs

Green and Black's Organic Chocolate
www.greenandblacks.com
    Organic fair trade chocolate products

A Quick Start to UltraMetabolism:
How to Make the Program More Powerful

UltraMetabolism is designed to distill the knowledge I've gained as a practitioner of Functional Medicine for the last 20 years into an easy-to-use format that allows you to easily lose weight, get healthy, and feel vitally alive—perhaps for the first time in your life. I refer to this state of vital health and optimal weight as UltraWellness, and it is something that is available to everyone right now, if they know how to attain it.

This book has been magnificently successful in helping people achieve that goal. To date, thousands of readers who've adopted the program have successfully taken control of their weight and significantly improved their lives.

However, the original version of the UltraMetabolism Prescription—the one outlined in the earlier chapters of this book—has one drawback. It isn't the exact same system I use in my practice, so it doesn't have exactly the same effects.

In this appendix, I want to give you an opportunity that readers of the original version of *UltraMetabolism* did not have. I want to show you how to use the exact same elimination diet I use in my practice. This will help you get a quick start to the UltraMetabolism Prescription, so you can lose weight even faster than you would following the original version of the program. It also promises to resolve more of your health symptoms more quickly than simply doing the program outlined in previous chapters.

In fact, this quick-start program may help you:

- Lose up to 7 pounds in five days
- Gain relief from many chronic symptoms rapidly and effectively (I will talk about exactly which symptoms might benefit a little later in this chapter)
- Enjoy a more powerful experience of health and weight loss more quickly than you would following the original UltraMetabolism Prescription
- Jumpstart your metabolism so you can move toward your optimum weight and health with greater ease
- Lose more weight, more rapidly—yet do so safely

I now have the chance to do what I always wanted to do with *UltraMetabolism*—create a powerful and comprehensive program that mimics what I do in my practice so you can achieve the same weight loss and health benefits that my clients do. This quick-start may give you an experience with health and weight loss you would not otherwise have—one so powerful that when it's over, you will never want to go back to your old ways of eating and living.

So how does it work?

Addressing Inflammation and Toxicity: The Keys to Automatic Weight Loss and Health

The quick-start program is founded on two fundamental principles of biology. To regain health and achieve consistent, permanent weight loss, you need to aggressively address the two major underlying causes of disease and obesity: *toxicity* and *inflammation*.

The same things that make people sick make them fat. In fact, being overweight is actually a symptom of an underlying health problem. And being fat creates even more sickness and disease—including heart disease, cancer, arthritis, dementia, and diabetes.

The answer to effective, long-term weight loss is to address the underlying causes of obesity and disease. To do this, you have to rebalance the 7 fundamental systems on which your entire physiology is based. When these systems are out of balance, they become the 7 roots of disease and obesity. When they are in balance, they serve as the 7 Keys to UltraMetabolism and UltraWellness.

I have already discussed this throughout the book, so you should be quite familiar with these concepts by now. However, what you might not realize is that if you're struggling with your weight, two of these keys are the primary cause of your problems: *toxicity* and *inflammation*.

Why?

To answer that question, let me give you a brief refresher course on toxicity and inflammation and the negative impact they have on your health. Remember, you can find more information on these topics in chapters 11 and 15.

Living in a Sea of Toxins: How Toxins Cause Obesity (and Other Illnesses)

I refer to toxins in the broadest sense—the sum total of our poor diet, chronic stress, and environmental pollutants that overload and poison our bodies and minds.

This includes toxic foods such as sugar, high-fructose corn syrup, trans fats, food additives and preservatives, pesticides, hormones and antibiotics in our food supply and water supply, as well as mercury, lead, and other heavy metals.

It also includes toxic thoughts, behaviors, and beliefs that keep us chronically stressed. These thoughts produce a flood of hormones, such as cortisol, that promote weight gain around the middle, increase our blood sugar and blood pressure, and even kill brain cells.

Exposure to toxins comes from two main sources: the environment (external toxins); and by-products of our metabolism and imbalances in our digestive system (products of our metabolism that are created when we break down food—internal toxins). Here is a list of some of the most common toxins you may be exposed to.

External Toxins: The Dangers from Without

- ❖ **Heavy metals.** Lead, mercury, cadmium, arsenic, nickel, and aluminum
- ❖ **Chemical toxins.** VOCs (volatile organic compounds— chemicals such as benzene and formaldehyde that are used to make everyday products like carpet or furniture and are slowly released from them over time), solvents (cleaning materials, formaldehyde, toluene, benzene), medications, alcohol, pesticides, herbicides, and food additives
- ❖ **Hidden infections.** Dental infections, hepatitis C virus, and mold toxins (sick building syndrome) are among the more common
- ❖ **Dietary toxins.** Sugar, high-fructose corn syrup (these are the two most common and prevalent causes of abnormal liver function—7 million Americans have a fatty liver from too much sugar), trans fatty acids, alcohol, caffeine, aspartame, pesticides, genetically modified foods, and the various plastics, pathogens (bugs), hormones, and antibiotics found in our food supply

Internal Toxins: Danger from Within

- ❖ **Microbial compounds.** Bacteria and yeast in the gut produce waste products, metabolic by-products, and cellular debris that can interfere with many of the body's functions and lead to increased inflammation and oxidative stress. These products include endotoxins, toxic amines, toxic derivatives of bile, and various carcinogenic substances such as putrescine and cadaverine

❖ **By-products of normal protein metabolism.** Urea, ammonia, and others require detoxification

So this is a lot of garbage to deal with. No wonder our systems become overloaded, and we get sick and fat!

I've seen it over and over with my patients, particularly those who get stuck and can't lose weight. They are toxic, and until they detoxify, they won't achieve their weight loss goals.

How Do Toxins Interfere with Weight and Metabolism?

Toxins affect your metabolism in several ways.

First, thyroid hormone function is altered by toxins. They cause your liver to excrete more of your active thyroid hormones, so your thyroid and metabolism slow down. They also damage the thyroid receptors, which are the docking stations for thyroid hormone.

Second, toxins interfere with appetite control mechanisms in the brain.

Third, toxins promote inflammation, which increases insulin resistance. This promotes fat storage and leptin (the "feeling full hormone") resistance, which in turn prevents your brain from recognizing that you are full.

Fourth, toxins disrupt energy production by your mitochondria, the little factories that turn food and oxygen into energy. There are thousands of mitochondria inside your cells, and when they are damaged, your metabolism grinds to a halt.

And lastly, the oxidative stress and free radicals created by toxins activate signals inside your cells to slow metabolism and increase insulin resistance.

Whew! No wonder our metabolism is jammed and not burning the calories and fat we put in our bodies.

This quick-start program is designed to help you detoxify, so you can begin to eliminate the factors that are standing in the way of vital health and weight loss. I will show you how that happens in a moment.

Before I do, let me explain the other reason the quick start is so powerful.

Food Allergies and Inflammation:
Keys to Your Weight and Health

We live an inflamed life. The sugar in our diets, high doses of the wrong oils and fats, hidden food allergens, lack of exercise, chronic stress, and hidden infections all trigger a raging, unseen inflammation deep in our cells and tissues that leads to every one of the major chronic diseases of aging—heart disease, cancer, diabetes, dementia, and more.

And it's by far the major contributor to obesity. Being inflamed makes you fat, and being fat makes you inflamed.

Being fat is being inflamed. Period!

If you don't address inflammation by eliminating hidden food allergens or sensitivities, and by eating an anti-inflammatory diet, you will likely not succeed at effective and permanent weight loss.

A key part of this quick-start program is helping you identify and avoid common food sensitivities or allergies. This is a huge part of the chronic health and weight problems so many suffer with. And though they are real and well documented in medical literature, they are generally ignored by conventional medicine.

What are food allergies anyway?

Conventional allergists and immunologists generally recognize only one kind of food reaction. This is the acute or type 1 hypersensitivity—the IgE mediated response—which turns on a histamine reaction.

It is a sudden, dramatic allergic response to things like peanuts or shellfish that can cause hives, trouble breathing, and even death within minutes. This is why conventionally trained allergists do skin or prick testing (PRT) or blood testing (RAST) to look for IgE antibodies to foods.

As important as it is to treat these kinds of allergies immediately and aggressively, practitioners of functional, integrative, and alternative medicine have long recognized the limitations of recognizing *only* these types of allergies. New research confirms what we have long known—that delayed reactions to food are controlled by a different part of the immune system (IgG antibodies and immune complexes).

Delayed reactions of this nature are called IgG delayed hypersensitivity reactions. This kind of reaction is much more common and creates much more suffering for millions of people than the acute allergies described above. However, it is mostly ignored by conventional medicine.

These delayed allergic reactions can cause symptoms anywhere from a few hours to 72 hours after eating, making it very difficult to connect the dots and see that what you just ate is connected to how you feel.

To complicate things even more, the symptoms are often vague.

Typical symptoms include fatigue, bloating, brain fog, food cravings, sinus congestion or post nasal drip, acne, eczema, psoriasis, irritable bowel, acid reflux, headaches, joint pain, trouble sleeping, weight gain, autoimmune diseases, asthma, and more.

Nonetheless, delayed food sensitivities play a huge role in many chronic illnesses and weight problems. In fact, treating food allergies and food sensitivities is one of the most beneficial things I do in my practice.

So what have I found after years of testing people for IgG allergies and teaching them how to use elimination diets (like the one you will learn in this quick-start program) to help them recover from their chronic symptoms and illnesses?

While everyone is different, there are some foods that irritate the immune system more than others. They are gluten (found in wheat, barley, rye, oats, spelt, triticale, and kamut); dairy (milk, cheese, butter, yogurt); corn; eggs; soy; nuts; nightshades (tomatoes, bell peppers,* potatoes, eggplant); citrus; and yeast (baker's and brewer's yeast, fermented products like vinegar).

These foods can also cause acute allergic reactions. Such reactions are rare, generally affecting less than 1 percent of the population. But when they occur, they are serious and permanent, and they need to be treated without delay.

For over 50 percent of us, however, there are some foods that just don't agree with us and take away from vibrant, good health.

How Do You Know If You Are Allergic or Sensitive to Foods?

There are two ways to find out if you are reacting to foods. One is a blood test for IgG antibodies to foods. This is useful and can pinpoint trouble areas, but it is not 100 percent accurate.

The second is a simple and well-accepted treatment called elimination/provocation. This means you get rid of the top trouble foods for 3–4 weeks, then reintroduce them one at a time and see what happens.

Eliminating any foods to which you are allergic is the basis for the remarkable results that people have shown when they follow the detox phase of the UltraMetabolism Prescription—losing weight, feeling better, and getting rid of chronic symptoms.

In the original version of the UltraMetabolism Prescription, I asked you to eliminate gluten, dairy, and eggs (along with a few other substances)

---

*Chili peppers have anti-inflammatory properties.

from your diet, because these are the top three foods that cause problems for most people.

With the quick-start program, you will eliminate an additional set of foods to which you may be sensitive for the first five days you are on the UltraMetabolism Prescription. Doing this offers you a stricter elimination diet than the one presented in chapters 20 and 21. You will see more dramatic health and weight loss results more quickly.

After these five days are over, you will carry on with Phase I of the UltraMetabolism Prescription. (As we go along, I will explain how to integrate the quick start with the UltraMetabolism Prescription so that they are one cohesive system.) You will be allowed to reintegrate some of the foods you eliminated during the quick start, so you can see if they are causing problems for you. This will give you the power to identify more precisely which foods are contributing to your health and weight problems, so you can make better choices about whether or not to keep eating these foods.

By following this elimination diet, people experience dramatic relief from many of the symptoms they thought they would have to live with for the rest of their lives. And they see a difference very quickly—often within five days. Eliminating and reintroducing foods this way not only connects the dots between what you eat and how you feel, it also allows you to repair your digestive and immune system, so you can be more resilient and tolerate a wider range of foods.

You see, most food sensitivities aren't permanent. That means the foods you love but have become sensitive to may not be a problem for you after you complete this five-day program. Even if they are, eliminating them for a total of twelve weeks usually allows your gut the chance to heal enough that you can eat them again from time to time. (I will explain how you can achieve these results in "Transitioning from the Quick Start into the UltraMetabolism Prescription" on page 351.)

In summary: using this quick start as you begin the UltraMetabolism Prescription may not only help you lose up to 7 pounds in 5 days, it also may provide you with a feeling of robust health and relief from many chronic symptoms very quickly.

It gives you an opportunity to achieve immediate weight loss. But more importantly, it may offer you renewed energy and relief from many chronic diseases.

And it helps you jump start your metabolism, setting the stage for an even more powerful experience during the remainder of the program.

How can you achieve such results in just five days?

Simple: Take away the things that make you toxic and inflamed; and provide your body with foods and activities that help you detoxify and cool inflammation.

Your body does the rest automatically. It has a natural ability to find balance and heal once you stop doing things that throw it off balance and provide things that put it back in balance.

As you eliminate the major sources of toxins in your life—addictive habits such as coffee, sugar, alcohol, processed food, fast food, junk food, trans fats, and high-fructose corn syrup—and reduce toxic stress for only five days, your body can renew and rejuvenate itself.

As you eliminate the major sources of inflammation in your diet—food allergens, sugar and flour products, and bad fats—your body can heal.

Then, as you eat whole, detoxifying, anti-inflammatory foods, the power of the program can take full effect.

What this promises is to supercharge your experience on the UltraMetabolism Prescription. It gives you a method for making the weight loss and health results you experience even more profound using one, simple five-day system. Here's how it happens.

The Power of the Quick-Start Program: How to Supercharge Your Experience with UltraMetabolism

In Phase I of the UltraMetabolism Prescription, you start by removing the three major foods that contribute to toxicity and inflammation: gluten, dairy, and eggs. I also ask you to remove a few other substances that lead to health and weight problems. However, the major premise of Phase I is to take gluten, dairy, and eggs out of your diet, so you can see if they are creating imbalances in your weight and health.*

As many problems as they create for some, gluten, dairy, and eggs aren't the only foods that cause people to become sick and fat. In fact, there are many other foods to which people can develop sensitivities.

---

*In addition to these foods, you are asked to eliminate various toxic substances like trans fats, high-fructose corn syrup, caffeine, alcohol, white sugar, white flour, and artificial sweeteners during the preparation week. That's because most of these substances are so poisonous to your biology that you shouldn't have been eating them in the first place. Unfortunately, they are so prevalent in our culture that most people don't even realize they are accidentally poisoning themselves by eating these foods. You will still have an opportunity to do the preparation week using the quick-start program, so you will be able to take advantage of its healing power.

In this quick-start program. I am going to ask you to eliminate *an additional set of foods that you may be sensitive to* as you begin the UltraMetabolism Prescription. I am also going to give you a specific daily diet regimen filled with whole foods that are naturally anti-inflammatory and detoxifying. You will:

- **Get rid of bad foods.** Eliminate foods that create toxicity and inflammation
- **Add good foods.** foods that are detoxifying and anti-inflammatory
- **Detoxify.** Drink a special, cleansing and detoxifying UltraBroth
- **Reduce inflammation.** Make delicious, anti-inflammatory UltraShakes
- **Relax.** Take a fabulously relaxing and detoxifying UltraBath every night before bed

You will do this for the first five days of Phase I on the UltraMetabolism program. A bit later in this appendix, I will explain exactly how and when to integrate this quick start into the program. I will also show you how to reintroduce, over the remainder of Phase I, the foods that you've systematically eliminated in the quick start. In this way, you can see which of these foods might be creating problems for you.

This regimen is stricter than the one outlined in chapters 21 and 22. It requires that you eliminate more foods from your diet, drink special shakes and broths, and take relaxing, detoxifying baths. But the benefits are worth the extra effort. You may lose weight and feel healthier more quickly; you will reset your metabolism much faster and more effectively; and if you're like many people, you may begin to experience some or all of the following benefits after the 5-day quick start:

- Better digestion and elimination
- Fewer symptoms of chronic illness
- Improved concentration, mental focus, and clarity
- Improved mood and increased sense of internal balance
- Increased energy and sense of well-being
- Less congestion and fewer allergic symptoms
- Less fluid retention
- Less joint pain
- Increased sense of peace and relaxation
- Enhanced sleep
- Improved skin

In addition, here is a partial list of the problems that can result from toxic overload and inflammation, and thus may be improved with this quick-start program:

- Bad breath
- Bloating, gas, constipation, and diarrhea
- Canker sores
- Difficulty concentrating
- Excess weight or difficulty losing weight
- Fatigue
- Fluid retention
- Food cravings
- Headaches
- Heartburn
- Joint pain
- Muscle aches
- Puffy eyes and dark circles under the eyes
- Sinus congestion
- Postnasal drip
- Skin rashes
- Sleep problems

Chronic problems with the body's ability to cleanse itself of toxins and reduce inflammation can also contribute to more serious conditions. If you suffer from any of the following, this program may help you fight them:

- Arthritis
- Autoimmune diseases
- Chronic fatigue syndrome
- Diabetes
- Fibromyalgia
- Food allergies
- Headaches
- Heart disease
- Inflammatory bowel disease (Crohn's or ulcerative colitis)
- Irritable bowel syndrome
- Menopausal symptoms (mood changes, sleep, hot flashes)
- Menstrual problems (premenstrual syndrome, heavy bleeding, cramps)

And you can achieve these results without:

- **Going hungry.** This is not one of those starvation diets. You shouldn't be hungry during the quick-start program, and you won't be as long as you follow my recommendations. In fact, one of the most common points of feedback that I've received from people who have been on this program has been how they actually had prepared too much food and didn't need it all since they simply weren't hungry.
- **Having cravings.** You're worried about cravings? Forget it! My patients have overwhelmingly reported that their cravings decrease when they do the program.
- **Eating unappealing foods.** You also don't have to worry about eating anything disgusting or tasteless. The options I offer are delicious.
- **Exercising too much.** Think you are going to have to spend two hours at the gym every day? Not so! Although I always recommend getting exercise on a daily basis, for just these five days, the eating program is powerful enough that you don't have to break a sweat if you don't want to.
- **Waiting too long.** Frequently people complain that it takes too long to see the benefits of any "diet." As a medically designed eating plan, we are working with the natural forces of your body, instead of against them like traditional diets do. As a consequence, you typically see the benefits very quickly.

I hope this will be the beginning of a new way of life for you. Not because I say it is good for you. No! Your life will change because you not only may lose weight, have more energy, and feel fantastic, you also may—probably for the first time in your life—have a taste of vital well-being, a taste of UltraWellness.

If you're wondering whether or not the more intense form of the program outlined in this quick-start guide will really be worthwhile for you, I think you might find the answer in the following quiz. It will help you determine just how toxic and inflamed you are. The more toxic and inflamed you are, the more likely it is that this quick-start program will impact your weight and health in *dramatic* ways, ways that the original UltraMetabolism Prescription may not achieve on its own.

How Toxic and Inflamed Are You?

Take the following quiz right now to see how much the program may help you. Write your answers in the Before column.

Then, to get a quantitative assessment for exactly how much change you experience, complete the quiz once more on the last evening of the 5-day quick start program. Write your answers in the After column. Then calculate the difference and see how much of a difference five short days can make in the quality of your health and life.

Please note that this questionnaire is not a replacement for regular checkups or medical diagnoses by your health care professional

TOXICITY AND INFLAMMATION QUIZ

Rating Scale

0 = Never or almost never have the symptom
1 = Occasionally have it, effect is not severe
2 = Occasionally have it, effect is severe
3 = Frequently have it, effect is not severe
4 = Frequently have it, effect is severe

Digestive Tract	Before	After	Difference
Nausea or vomiting			
Diarrhea			
Constipation			
Bloating			
Belching or passing gas			
Heartburn			
Intestinal/stomach pain			
TOTAL			

Ears	Before	After	Difference
Itchy ears			
Earaches or ear infections			
Drainage from ear			
Ringing in ears or hearing loss			
TOTAL			

Emotions	Before	After	Difference
Mood swings			
Anxiety, fear, or nervousness			
Anger, irritability, or aggressiveness			
Depression			
TOTAL			

(continued)

Energy/Activity	Before	After	Difference
Fatigue or sluggishness			
Apathy or lethargy			
Hyperactivity			
Restlessness			
TOTAL			

Eyes	Before	After	Difference
Watery or itchy eyes			
Swollen, reddened, or sticky eyelids			
Bags or dark circles under eyes			
Blurred or tunnel vision (does not include near- or far-sightedness)			
TOTAL			

Head	Before	After	Difference
Headaches			
Faintness			
Dizziness			
Insomnia			
TOTAL			

Heart	Before	After	Difference
Irregular or skipped heartbeat			
Rapid or pounding heartbeat			
Chest pain			
TOTAL			

Joints/Muscles	Before	After	Difference
Aches or pain in joints			
Arthritis			
Stiffness or limitation of movement			
Aches or pain in muscles			
Feeling of weaknessor tiredness			
TOTAL			

Lungs	Before	After	Difference
Chest congestion			
Asthma or bronchitis			
Shortness of breath			
Difficulty breathing			
TOTAL			

Mind	Before	After	Difference
Poor memory			
Confusion or poor comprehension			
Poor concentration			
Poor physical coordination			
Difficulty making decisions			
Stuttering or stammering			
Slurred speech			
Learning disabilities			
TOTAL			

Mouth/Throat	Before	After	Difference
Chronic coughing			
Gagging or frequent need to clear throat			
Sore throat, hoarseness, or loss of voice			
Swollen or discolored tongue, gum, or lips			
Canker sores			
TOTAL			

(continued)

Nose	Before	After	Difference
Stuffy nose			
Sinus problems			
Hay fever			
Sneezing attacks			
Excessive mucus formation			
TOTAL			

Skin	Before	After	Difference
Acne			
Hives, rashes, or dry skin			
Hair loss			
Flushing or hot flushes			
Excessive sweating			
TOTAL			

Weight	Before	After	Difference
Binge eating/drinking			
Craving certain foods			
Excessive weight			
Compulsive eating			
Water retention			
Skipping meals often			
Excess alcohol intake			
Night eating			
TOTAL			

Other	Before	After	Difference
Frequent illness			
Frequent or urgent urination			
Genital itching or discharge			
TOTAL			
GRAND TOTAL			

Key to the Questionnaire

Add your individual scores for each group of symptoms. Add your group scores for a grand total. Then check the chart below to assess the level of your health problem and the potential benefits of the quick-start program.

Your Score	Health Status	Benefits You May Receive*
10 or less	Optimal health	Increased energy, improved mood, and weight loss
11–50	Mild imbalance	In addition to the above, you may see improved digestion, better skin, and less nasal congestion.
51–100	Moderate imbalance	You may experience all of the above as well as reduced joint pain, muscle aches, headaches, and more.
Over 100	Severe imbalance	You may experience much of the above, but to deeply address your chronic symptoms, you will need the support of a physician trained in functional medicine.

*The benefits for each progressive level of health imbalance are additive. In other words, the more imbalanced your health currently is, the more benefits you are likely to receive from the program. For example, if you currently have a moderate health imbalance (a score of 51–100), you may experience all of the benefits listed for earlier categories (increased energy, improved mood, weight loss, improved digestion, better skin, and less nasal congestion) as well as the benefits listed in your category.

Now that you know how toxic and inflamed you are, it's time to get a quick start to UltraMetabolism. Let me explain how to integrate this system into the eight-week UltraMetabolism Prescription, so you can achieve optimal health and weight loss results.

How to Get a Quick Start to UltraMetabolism

The foundation of the quick-start program is actually quite simple.

❖ Avoid the foods and substances that cause toxicity and inflammation.
❖ Enjoy detoxifying and anti-inflammatory foods.

✢ You should not be hungry. Be sure to eat enough food
   and shakes to feel gently satisfied but not full.
✢ Your body does the rest.

Here are the basic principles—what foods to avoid, which ones to
enjoy, a shopping list, guidelines for integrating the quick start into the
UltraMetabolism Prescription presented in chapters 20 and 21, and a plan
for each of the five days of the quick start.

Foods and Substances You'll Avoid for Five Days

During the quick start, you will avoid the following foods. You will do this
effortlessly, simply by sticking with the program instead. In other words,
you don't have to worry about *not* eating these foods as long as you follow the guidelines set forth in the program. They've already eliminated the
potentially offending foods.

The program is designed to take all the thinking and choosing out
of your life for five days so you can experience its dramatic benefits.

The list below is just to remind you of the toxic and inflammatory
foods that most of us consume on a daily basis.

Some of the fruits and vegetables, and even the eggs and meats mentioned here, may not be a problem for you. The only way to find out,
however, is to give up all of them for five days and listen to what your
body tells you.

Do you feel better? Do you get sick when you eat them? Pay attention. Your body has much to teach you if you listen.

✢ Sugar (white sugar, cane sugar, dehydrated cane juice, brown
   sugar, honey, maple syrup, high-fructose corn syrup, sucrose,
   glucose, maltose, dextrose, lactose, corn syrup, and white grape
   juice concentrate)
✢ Sugar alcohols such as sorbitol, mannitol, xylitol, and maltitol
✢ Artificial sweeteners such as aspartame
✢ Natural sweeteners such as stevia (although it may be fine in the
   long run, in the short run it stimulates sweet cravings and will
   sabotage your efforts)
✢ Alcohol
✢ Caffeine (coffee, teas except for green tea, sodas)
✢ Citrus fruits and juices (except for lemon juice, unless you are al-
   lergic to it)
✢ Yeast (baker's and brewer's, fermented foods like vinegar)★
✢ Dairy products (milk, butter, yogurt, cheese)★

❖ Eggs★
❖ Gluten★ (hidden in many foods, including anything containing wheat, barley, rye, spelt, kamut, triticale, or most oats; see www. celiac.com for a comprehensive list of gluten-containing foods)
❖ Corn★
❖ Beef, pork, lamb, or other mea,t except organic poultry
❖ Nightshades (tomatoes, potatoes, eggplant, bell peppers)
❖ Peanuts★
❖ Refined oils and hydrogenated fats such as margarine, and almost every processed food in the supermarket (even foods labeled as "trans fat free" can have 0.5 gram per serving, according to new government regulations)
❖ Stimulants (including decongestants, diet pills, ephedra, ma huang, and yerba maté)
❖ All flour products
❖ Processed foods or food additives (check labels carefully!)
❖ Fast food
❖ Junk food
❖ Any food that comes in a box, package, or can or is otherwise commercially prepared and filled with chemicals, preservatives, and other unnatural ingredients to make it shelf-stable

Foods You Will Enjoy

Here are the foods you will eat during the 5-day quick start:

❖ Filtered water (six-eight glasses per day)
❖ Fish, especially small, non-predatory species such as sardines, herring, wild salmon, black cod or sable fish, sole, and cod
❖ Lean white meat chicken breast (preferably organic)
❖ Fresh or frozen non-citrus fruits, ideally only berries (preferably organic)
❖ Fresh vegetables (preferably organic)
❖ Fresh vegetable broth (3-4 cups per day)
❖ Legumes (lentils, navy beans, adzuki beans, mung beans, tofu, and others)

★The most common food allergens are dairy, gluten, eggs, corn, yeast, and peanuts. Soy and nuts can be allergens for some people, but they have so many benefits that I include them in my program. If you are sensitive to soy or particular nuts, replace them with other beans, nuts, seeds, fish, or lean poultry.

❖ Brown rice

❖ Nuts and seeds (almonds, walnuts, pecans, macadamia nuts, and
   pumpkin seeds)

❖ Flaxseeds (ground, preferably organic)

❖ Lemons (preferably organic—don't buy pre-squeezed lemon juice)

The Quick Start Program

Now that you know which foods to avoid and which to enjoy, I want
to introduce you to the basic meal plan you will follow every day for
five days.

Follow the plan as best you can. It is important to have the snacks
and broth in between. It will keep you feeling balanced and satisfied. You
can even buy a thermos to keep the broth hot all day!

That doesn't mean you *have* to eat every single thing listed below.
Don't stuff yourself. There is so much food on this program that the only
danger you face is making yourself too full! Don't overdo it. Just eat until
you are gently satisfied at each meal.

(Please note, recipes for the UltraBroth, UltraShake, and UltraBath
follow on pages 341–344. For cooking tips and helpful preparation
instructions for some of the other items on the quick-start program, see
chapter 17.)

The 5-Day Quick Start

Breakfast (7–9 am): 1 cup of decaf or caffeinated green tea
   steeped in hot water for 5 minutes (you may also have the
   green tea later in the day; limit your intake to 2 cups a day);
   UltraShake

Morning Snack (10–11 am): 1 cup UltraBroth; Another
   UltraShake (if you are hungry)

Lunch* (12–1 pm): 2 cups or more of steamed or lightly
   sautéed veggies (eat enough to feel gently satisfied); ½ cup
   cooked brown rice; ½ cup fruit or berries for dessert (either
   here or at dinner, not both, and only 1–2 times during the
   5-day program); UltraShake (optional)

Afternoon Snack (2–3 pm): 1 cup UltraBroth; UltraShake
   (if you are hungry)

---

*Note: You can switch the lunch and dinner menus, if you prefer.

Dinner (5–7 pm): 4–6 ounces of fish or chicken breast cooked with olive oil and lemon juice *or* 4–6 ounces of tofu or legumes† (canned legumes are okay, but be sure to rinse them well). I encourage you to spice up your meals with rosemary, cilantro, ginger, garlic, turmeric, or curry, and sea salt; 2 cups or more of steamed or lightly sautéed veggies (eat enough to feel gently satisfied); ½ cup cooked brown rice; 1 cup UltraBroth

How to Integrate the 5-Day Quick Start into the 8-Week UltraMetabolism Program

Incorporating the 5-day quick start into the Ultrametabolism Prescription is easy. Here's what you need to do:

1. Start by doing the preparation week as outlined in chapter 19. This week will get your body ready for all the goodness to come, and will help you eliminate toxic substances from your diet in a systematic way to make the transition into the quick start simple and painless.
2. Follow the program outlined above for the first 5 days of Phase I. Do this instead of starting Phase I as prescribed in chapter 20.
3. As the 5-day quick start comes to a close, you are going to move to Phase I of the UltraMetabolism Prescription. That means you will have 2 days left in the first week of Phase I, and an additional 2 weeks to go in that stage of the program.

    You will make this switch by following the food reintegration guidelines presented below in "Transitioning from the Quick Start into the UltraMetabolism Prescription." At this time, you can use the recipes, daily menus, and dietary recommendations outlined in chapter 20, but be careful to check for foods you may not have systematically reintegrated yet. (See on page 351 for more details.)
4. When Phase I comes to a close, proceed to Phase II as recommended in chapter 21.
5. Stay on Phase II for a lifetime of UltraMetabolism and Ultra-Wellness.

---

†Beans can be prepared quickly and easily by opening a can of white cannelli beans and adding it to ½ cup of chopped onions sautéed in 1–2 tablespoons extra virgin olive oil and fresh rosemary. Season with a pinch of sea salt and you have a delicious Tuscan bean dish in minutes.

Keep in mind that you can adopt all of the other parts of the UltraMetabolism Prescription described in the rest of the book, if you choose to, while you complete the 5-day quick start. You can add exercise, supplements, and relaxation strategies; you can work on optimizing the keys during these five days by following the guidelines in part II; or you can do the 5-day quick start first and then follow up with other components of the UltraMetabolism program as you move into Phases I and II. The choice is yours.

Either way, the premise is simple. Follow the 5-day quick-start outlined above in place of the first five days of Phase I of the UltraMetabolism Prescription. Then move into the remainder of Phase I, reintegrating foods as explained later in this appendix. When you complete Phase I, move into Phase II as outlined in chapter 21. Using this system, you will be well on your way to a lifetime of UltraWellness!

Helpful Tips During the 5-Day Quick Start

- Try to begin the quick-start on a Sunday (and the preparation week on the previous Saturday) to give yourself time to acclimatize to the program. This is very important.
- Try to eat or drink something every one to two hours during the day to maintain energy and well-being.
- Take the UltraBroth any time you feel weak or hungry.
- Don't worry if you feel foggy or fatigued or you suffer from other withdrawal symptoms for the first day or two. This is a good sign! The foods we are allergic to also tend to be the foods we are most addicted to. Getting rid of these foods sometimes causes withdrawal symptoms. It means your body is detoxifying, and you are on your way to UltraWellness. Your symptoms should clear up quickly.
- If you do feel tired, rest or take a nap. Take care of yourself during this time.
- Make sure you are drinking plenty of water—six to eight glasses a day. Besides helping you stay hydrated, this happens to be one of the best ways to keep withdrawal symptoms at bay.
- Make sure you are having at least one or two bowel movements a day; otherwise you may feel more toxic, with fatigue, sluggishness, brain fog, headaches, bloating, and other symptoms. If you are not moving your bowels daily, then try taking 1,000–2,000 mg of buffered vitamin C twice a day, along with 100–200 mg of magnesium citrate once or twice a day.

The UltraBroth Recipe

On the 5-day quick start, I suggest drinking 3–4 cups of vegetable broth per day. It is very similar to Dr. Hyman's Detox Broth in chapter 20. The broth is a wonderful, filling snack that will also provide you with many healing nutrients and alkalinize your system, making it easier to detoxify, lose weight, and feel great.

Our modern diet—with its sugar, processed foods and excess animal protein—is an acid-producing diet. This sort of diet creates a toxic cellular environment. Our cells function optimally in a slightly alkaline environment.

Many modern diseases are related to excess acidity in the body. Detoxification can happen only if we reduce the acidity in our bodies. The Ultra-Broth is a simple way to alkalinize your body. It can be enjoyed during the 5-day quick start and beyond.

The following recipe can be varied according to taste.

For every 3 quarts of water, add:
1 large chopped onion
2 sliced carrots
1 cup daikon or white radish root and tops (ideal, but optional)
1 cup winter squash, cut in large cubes
1 cup root vegetables such as turnips, parsnips, and rutabagas for
     sweetness
2 cups chopped greens: kale, parsley, beet greens, collard greens, chard,
     dandelion, cilantro or other greens
2 celery stalks
½ cup seaweed such as nori, dulse, wakame, kelp, or kombu
½ cup cabbage
4½-inch slices fresh ginger
2 cloves whole garlic (not chopped or crushed)
Sea salt to taste
1 cup fresh or dried shitake or maitake mushrooms, if available (these have
     powerful immune boosting properties)

Add all of the ingredients at once and place on a low boil for approximately 60 minutes. It may take a little longer. Simply continue to boil to taste.

Cool, strain (throw out the cooked vegetables), and store in a large, tightly sealed glass container in the fridge.

Simply reheat as needed and drink at least 3–4 cups a day.

A few notes about the broth: After making the broth, you may discard the vegetables, as they are not intended to be eaten (but be sure to keep the broth!). Feel free to mix, match, and vary the vegetables to create your own version of the broth. You need to wash the vegetables well, scrubbing the root vegetables with a vegetable brush, but you don't need to peel them. Cut them into large chunks, so you can fit them in the pot. If you just can't make the broth, you can substitute low-sodium, organic vegetable broth from Pacific Foods or Imagine Foods, though it is second best to fresh broth.

The UltraShake Recipe

This shake provides essential protein for detoxification, omega-3 fatty acids from flax oil, fiber for healthy digestion, flaxseeds for increased elimination, and antioxidants and phytonutrients from the berries and fruit. It will sustain you, even out your blood sugar, and help control your appetite throughout the day.

For better-tasting shakes: Adding frozen cherries, ½ of a frozen banana, and nut butter provides the best tasting shake.

Version 1: Using Rice Protein

This shake is the easiest to make and digest—and quite satisfying, too.

> 2 scoops rice protein powder (this is the average amount; follow the directions for the serving size of the product you choose)
> 1 tablespoon organic, combination flax and borage oil
> 2 tablespoons ground flaxseeds
> Ice (made from filtered water), if desired
> 6–8 ounces filtered water to desired consistency (some like thicker drinks, some thinner)
> ½ cup frozen or fresh, non-citrus organic fruit such as cherries, blueberries, raspberries, strawberries, peaches, pears, or frozen bananas
> 1 tablespoon nut butter (almond, macadamia, pecan) or ¼ cup nuts (almonds, walnuts, pecans, or any combination of these), soaked overnight (optional)

Special note on rice protein: I prefer detoxifying hypoallergenic rice protein. While it can be expensive, it replaces meals and facilitates detoxification and weight loss. Please check with your local health food store for a good brand.

Note: Use the flaxseeds in up to 2 shakes a day, no more.

Version 2: Fruit and Nut Smoothie

If you don't want to use rice protein, you can simply substitute silken tofu.
This is a nice creamy, shake made from real food.

¼ cup drained silken tofu

½ cup plain, unsweetened, gluten-free soy milk (such as Silk)

1 tablespoon organic, combination flax and borage oil

2 tablespoons ground flaxseeds

½ cup fresh or frozen, non-citrus organic fruit such as cherries, blueberries,
    raspberries, strawberries, peaches, pears, or frozen bananas

1 tablespoon nut butter (almond, macadamia, pecan) or ¼ cup nuts (almonds,
    walnuts, pecans, or any combination of these), soaked overnight (optional)

Ice (made from filtered water), if desired

2–4 ounces filtered water to desired consistency (some like thicker drinks,
    some thinner)

Note: Use the flaxseeds in up to 2 shakes a day, no more.

Version 3: Nut Smoothie

This shake is designed to be soy free. It requires no extra purchase of powder and can be made from easily accessible ingredients.

½ cup plain, unsweetened, gluten-free almond or hazelnut milk

1–2 tablespoons nut butter (almond, macadamia, pecan) or ¼ cup nuts
    (almonds, walnuts, pecans, or any combination of these), soaked overnight

1 tablespoon organic, combination flax and borage oil

2 tablespoons ground flaxseeds

½ cup fresh or frozen, non-citrus organic fruit such as cherries, blueberries,
    raspberries, strawberries, peaches, pears, or frozen bananas

Ice (made from filtered water), if desired

2–4 ounces filtered water to desired consistency (some like thicker drinks,
    some thinner)

Note: Use the flaxseeds in up to 2 shakes a day, no more.

The UltraBath

The UltraBath is a key component of the 5-day quick start. It provides many powerful benefits in one easy, 20-minute solution every day. It may not seem important, but the UltraBath has become a favorite feature of the quick-start program among those who have followed it.

The benefits of the UltraBath may include:

- Relaxation of your nervous system and a reduction in cortisol through the use of lavender oil, which promotes weight loss and lessens inflammation
- Enhancement of detoxification through the effects of the magnesium and sulfur in Epsom salts
- Enhanced sleep through the effects of the hot bath and magnesium
- Alkalinization of your body through the use of baking soda (sodium bicarbonate), which promotes an ideal pH for healing, detoxification, and optimal cellular function
- Increased circulation and heart rate, which serves as a form of passive exercise
- Lower blood pressure and blood sugar levels
- Increased heart rate variability, a sign of a healthy nervous system and reduced stress
- Increased sweating and elimination of toxins

Add 2 cups Epsom salts, 1 cup baking soda, and 10 drops lavender oil to bathwater as hot as you can tolerate.

Take a 20-minute UltraBath just before bed every night.

For extra-powerful detoxification, wrap yourself in towels immediately after the bath, get in bed under the covers, and sweat for 20 minutes more, then remove the towels before going to sleep. You can go directly to sleep without rinsing off after the bath.

You can also take a sauna or steam bath for up to 30 minutes per day, if it is available to you either in addition to your UltraBath or in place of it.

Shopping List

The quantities on this checklist should take care of all your needs for the 5-day quick start. Choose organic whenever possible.

Protein

You can choose from any combination of the following. Amounts will vary depending on how much of each you eat.

Fish—small, non-predatory species such as sardines, herring, wild salmon, black cod, sable fish, sole, and cod	1¾–2¾ pounds of fish and chicken combined
Boneless, skinless, chicken breast (preferably organic)	1¾–2¾ pounds of fish and chicken combined
Tofu	3 pounds
Canned beans: cannelli beans, navy beans, or chick peas	4 cans

Vegetables

You will need lots of these!

Choose a variety from each category identified on page 349 in A Special Note on What Vegetables to Buy. You should have enough to steam 2 cups each for lunch and dinner and to make the UltraBroth (see recipe on page 341). You can eat as many vegetables as you like. Buy enough so you're not hungry.	½–2 pounds per meal depending on your appetite

Whole Grains

Brown rice, long or short grain	4 cups

Oil

Extra virgin olive oil	1 liter

Beverages

Filtered or distilled water, or purchase a reverse osmosis or Brita filter to purify your water	Approximately 7 gallons, or enough for 8–10 glasses per day
Green tea, preferably decaf or caffeinated	1 box of tea or 8 ounces of loose leaf

Spices

These spices have powerful anti-inflammatory and detoxifying properties, which is why I recommend them for the quick start. You can use them to your personal tastes.

Fresh ginger	1 large root (4 ounces)
Whole garlic cloves	2 heads (or you can purchase pre-peeled cloves for convenience)
Lemons (Though they aren't a spice, you can still use them to flavor your food.)	8–12
Turmeric (the yellow spice found in curry—add 1–2 teaspoons to the cooking water for your rice)	1 small bottle
Rosemary (fresh is best)	1 bunch
Cilantro (also known as coriander—fresh is best)	1–2 bunches
Chili peppers (fresh is best)	1–2 peppers goes a long way!
Whole black peppercorns for the pepper mill	1 bottle
Sea salt★	1 bottle

★Use only sea salt during the program. Table salt is mined from underground salt deposits and includes a small portion of calcium silicate, an anti-caking agent added to prevent clumping. Because of its fine grain, a single teaspoon of table salt contains more salt than a tablespoon of kosher or sea salt. Sea salt is harvested from evaporated seawater and undergoes little or no processing. It contains nearly 50 minerals that support your health.

UltraShake Version 1

Rice protein powder	1 large bottle
Fresh or frozen non-citrus fruit such as cherries, blueberries, blackberries, or strawberries (this will also be enough to eat for dessert once or twice during the 5-day program)	4–6 cups or 2-3 packages Cascadian Farms organic fruit
Ground flaxseeds: You can buy flaxseeds already ground, or you can buy them whole and grind them yourself in a coffee grinder. Fiproflax is the freshest organic ground flax on the market. Be sure to keep it refrigerated.	1 15–ounce package or the equivalent in whole, bulk flaxseed
Combination flax and borage oil: organic, high lignan. Barlean's and Spectrum are the best brands.	1 large bottle
Nuts and seeds: almonds, walnuts, pecans, macadamia nuts, pumpkin seeds (optional)	2 cups
Nut butter: almond, macadamia, or pecan. You may substitute nuts for nut butter, if you prefer.	1 jar

UltraShake Version 2

Silken tofu (Note: This amount is in addition to the amounts of tofu recommended for your meals above.)	7 cups
Unsweetened, gluten-free soy milk (such as Silk)	14 cups
Fresh or frozen non-citrus fruit such as cherries, blueberries, blackberries, or strawberries (this will also be enough to eat for dessert once or twice during the 5-day program)	4–6 cups or 2-3 packages Cascadian Farms organic fruit

Ground flaxseeds: You can buy flaxseeds already ground, or you can buy them whole and grind them yourself in a coffee grinder. Fiproflax is the freshest organic ground flax on the market. Be sure to keep it refrigerated.	1 15–ounce package or the equivalent in whole, bulk flax seed
Combination flax and borage oil: organic, high lignan. Barlean's and Spectrum are the best brands.	1 12–ounce bottle
Nuts and seeds: almonds, walnuts, pecans, macadamia nuts, and pumpkin seeds (optional)	2 cups
Nut butter: almond, macadamia, or pecan. You may substitute nuts for nut butter, if you prefer.	1 jar

UltraShake Version 3

Plain, unsweetened almond or hazelnut milk	14 cups
Nuts and seeds: almonds, walnuts, pecans, macadamia nuts, and pumpkin seeds	2 cups
Nut butter: almond, macadamia, or pecan. You may substitute nuts for nut butter, if you prefer.	1 jar
Fresh or frozen non-citrus fruit such as cherries, blueberries, blackberries, or strawberries (this will also be enough to eat for dessert once or twice during the 5-day program)	4–6 cups or 2-3 packages Cascadian Farms organic fruit
Combination flax and borage oil: organic, high lignan. Barlean's and Spectrum are the best brands.	1 12–ounce bottle
Ground flaxseeds: You can buy flaxseeds already ground, or you can buy them whole and grind them yourself in a coffee grinder. Fiproflax is the freshest organic ground flax on the market. Be sure to keep it refrigerated.	1 15–ounce package or the equivalent in whole, bulk flax seed

UltraBroth

Quantities need to be multiplied by 3–4 batches, depending how much broth you consume per day. Note that the following amounts are in addition to the recommendations above.

Onion	1 large
Carrots	2
Daikon or white radish root and tops (ideal, but optional)	1 cup
Winter squash cut in large cubes	1 cup
Root vegetables: turnips, parsnips, and rutabagas for sweetness	1 cup
Greens: kale, parsley, beet greens, collard greens, chard, dandelion, cilantro, and/or other greens	2 cups or 1 bunch
Celery	2 stalks
Sea weed: nori, dulse, wakame, kelp, or kombu★★	½ cup
Cabbage	½ cup
Fresh ginger root	4½-inch slices
Whole garlic (not chopped or crushed)	2 cloves
Fresh or dried shitake or maitake mushrooms, if available (these have powerful immune boosting properties)	1 cup

UltraBath

Note: Amounts may vary depending on what your local store sells.

Baking soda	1 large box or 2 liters
Epsom salts	4–6½-gallon containers
Lavender essential oil	1 small bottle

★★Seaweed is a new food for most people. It is purchased dry in packages and simply needs to be broken off, measured, and thrown in the broth.

A Special Note on What Vegetables to Buy

During the 5-day quick start, you may choose from the following vegetables at lunch or dinner. These veggies maximize phytonutrient intake (phytonutrients are powerful disease fighting chemicals found in colorful plant foods). They also deliver potent detoxifying and anti-inflammatory compounds.

Choose vegetables from all of the categories. Use vegetables you like already, but also experiment and try something new!

I encourage you to eat primarily cooked vegetables (either steamed or sautéed) during the quick-start program, because they are easier to digest. While you may still enjoy salads or raw vegetables, keep them to a minimum until you have completed the quick start. You can eat more salads once you move into Phase I of UltraMetabolism. Use extra virgin olive oil, lemon juice, and salt and pepper for dressing if you choose to have a salad during these five days.

- **Allium vegetables.** Garlic, onions, leeks, and shallots
- **Cruciferous vegetables.** Broccoli, cabbage, kale, collard greens, kohlrabi, Brussel sprouts, bok choy, and Chinese broccoli
- **Dark blue, purple, or red fruits and vegetables.** Cherries, blueberries, blackberries, beets, red onions, purple grapes, red cabbage, and radicchio‡
- **Dark green, leafy vegetables.** Spinach, watercress, arugula, collard greens, kale, cabbage, Brussel sprouts, and loose-leaf lettuce (not iceberg lettuce)
- **Red and yellow vegetables.** Chili peppers, sweet potatoes (a different plant family than regular potatoes), winter squash (like acorn, butternut, buttercup, and kabocha), and carrots
- **Special detoxifying vegetables.** Artichokes or artichoke hearts, asparagus, beets, celery, and dandelion greens
- **Sea vegetables.** Nori, kelp, dulse, kombu, hijiki, arame, and wakame
- **Root vegetables for broth.** Sweet potatoes, rutabagas, turnips, parsnips, and carrots

That is all you need to know to get a powerful quick start to UltraMetabolism. Follow these guidelines, and you will have an even more intense experience with health and weight loss than you otherwise would have with the UltraMetabolism Prescription.

---

‡Please use the berries only in the shake and ½ cup as a treat a few days during the diet. They are not to be eaten in unlimited quantities.

Once you have completed the 5-day quick start, you will need to move into Phase I of the UltraMetabolism program. Let me explain how to do that now.

Transitioning from the Quick Start into the UltraMetabolism Prescription

Once you complete the 5-day quick start, the first thing you should do is retake the quiz on page 331. Compare your answers now with the ones you provided before you began the quick start. You will soon see the magnitude of your newfound vitality and health. This should inspire you to continue with the UltraMetabolism program on the path to lifelong, vital health.

Most importantly, you should be feeling better and losing weight. That is the real reward of the quick-start program.

However, your journey to UltraWellness and an UltraMetabolism has only begun. You have the opportunity to complete the remainder of the UltraMetabolism program to help you deepen and sustain the changes in health and weight you are experiencing. By doing so, you will lose even more weight and feel even better than you do right now.

That extra benefit is worth a little bit of extra effort.

As a step on the path toward UltraWellness, I recommend taking the time to systematically reintegrate the foods you eliminated during the 5-day quick start. In fact, *the single biggest gift you can give yourself throughout the rest of the UltraMetabolism Prescription is identifying which foods you are allergic to and which you can safely eat and enjoy.*

Carefully reintroducing foods as you move out of the 5-day quick start and into Phase I (and then again as you move out of Phase I and into Phase II) can be a very useful and educational experience. If you do it properly, you will discover which foods have been making you sick and fat. You can then keep them out of your diet and choose foods that make you thrive instead. Later on, you can even expand your diet to include items you have eliminated.

Some of this systematic reintegration will happen naturally as you enter Phase I. You will automatically be keeping certain foods out of your diet during the remainder of Phase I to allow your body additional time to detoxify and heal.

At this stage, however, you can begin reintegrating certain foods you eliminated during the 5-day quick start. And you can eat a little more liberally. As you begin this process, I recommend proceeding slowly, so you can take full advantage of what you have learned in the last five days.

The remainder of this section will focus on a step-by-step plan for reintroducing foods into your diet as you move into Phase I. I strongly encourage you to follow this plan. It will help you maintain the vital health you have worked so hard to achieve in the last five days and expand your understanding of how food works with your body.

Reintroduction of Potentially Allergenic Foods

Once you've taken the quiz, your next step is to start reintroducing some of the foods you eliminated during the 5-day quick start to see if any of them lead to symptoms or other problems. This will allow you to identify any offending foods and avoid them for a longer period of time, so your immune system can cool off a little more.

Add only one new food group every three to four days until you have reinstated all of the foods that you eliminated for the 5-day quick start but that are allowed in Phase I (I will give you a detailed list on the next page to simplify things). Doing this step by step will allow you to identify the foods within this list that are troublesome for you.

When you reintroduce these potentially allergenic foods, eat them at least three to four times a day for three days to see if you notice a reaction—unless, of course, you notice a problem right away. Then stop eating the food immediately.

Keep a log of any symptoms that you experience as you reintroduce different food groups. Symptoms may occur anywhere from a few minutes to 72 hours after ingestion and can include post nasal drip; digestive problems such as bloating, gas, constipation, and diarrhea; reflux; headaches; joint pain; fluid retention; fatigue; brain fog; mood changes; changes in sleep patterns; skin rashes; and more.

Tracking your symptoms should guide you to which foods trigger allergic reactions in your system. When you identify the foods to which you are allergic, it's best to avoid them for 90 days. Then you can reintroduce them again, but eat them no more than every four or five days.

Avoiding the foods to which you are sensitive for this additional time will give your immune system a chance to cool off. Then you may be able to eat many previously problematic foods once again, as often as every four to five days, while still keeping inflammation at bay. This will help minimize the health and weight problems these foods may otherwise cause you. If you eat them every day, you may become sensitive to them again.

You may find that you remain extremely sensitive to certain foods, even after your immune system's cooling-off period. In this case, you will have to choose what you want to eat and what you want to avoid. Put another way, you will be choosing between vital health and foods that may taste good but don't make you feel good.

If you decide to eat foods to which you are sensitive, you may want to do it only every once in a while. This will help minimize their impact on your weight and health.

Using a food log to track your symptoms and monitor your progress is an excellent way to identify what foods you can tolerate and what foods you are allergic to. There is not enough room here to include a sample; however, I have provided one in the downloadable *UltraMetabolism Companion Guide* at www.ultrametabolism.com/guide. It is a very detailed food log that you can use to fill in the blanks quite easily.

Here is a plan for how to reintroduce foods over the remainder of Phase I, so you can maximize the benefits of what you have just experienced with the 5-day quick start.

Food Reintroduction after Quick Start—Remainder of Phase I

Foods or Ingredients to Permanently Avoid
(See the list in chapter 20 for a more detailed account.)

+ High-fructose corn syrup
+ Trans or hydrogenated fats
+ Processed and junk foods
+ Fast foods
+ Artificial sweeteners
+ Foods with additives preservatives and colors

Continue to Avoid throughout the Remainder of Phase I
(See the list in chapter 20 for a more detailed account.)

+ Dairy (milk, cheese, butter, yogurt)
+ Gluten (barley, rye, oats, spelt, kamut, triticale, wheat—see www.celiac.com for a complete list of foods that contain gluten)
+ Eggs
+ Peanuts
+ Caffeine
+ Alcohol
+ Sugar in any form (table sugar, honey, maple syrup, corn syrup)

Foods to Reintegrate as You Enter Phase I

Generally, the following are problem-free foods for most people, so no reintegration period is necessary. Start enjoying them right away as you make the transition from the 5-day quick start.

- Fresh fruit (including citrus fruits)
- Raw vegetables, like salads; include artichokes, avocados, and olives
- Organic lamb, although you should keep these dense animal proteins to a minimum and consume no more than 4 ounces at a time (you can find organic lamb, beef, and pork at Whole Foods Markets or online)
- Low-allergy grains such as quinoa, buckwheat, and millet
- Healthy oils such as cold-pressed nut or seed oils, in addition to the extra virgin olive oil you have been using
- Any spices, in addition to the garlic, ginger, curry or turmeric, rosemary, and fresh cilantro you have been using

Besides gluten, dairy, and eggs, these are the most allergenic foods in your diet. Reintroduce them one at a time, with 3 days in between to monitor symptoms.

- Yeasted products (baker's and brewer's, fermented foods like vinegar)
- Soy and other foods not on the list of foods to avoid you may have eliminated during the quick start to test for food sensitivities
- Corn
- Nightshades

How to Use the Daily Menus with the Reintroduction Program

As you reintegrate foods into your diet and move into Phase I of UltraMetabolism, you can use the daily menus, recipes, and shopping lists provided earlier in the book.

Just be sure to check for ingredients you may not have reintegrated yet. For example, a few of the recipes in Phase I contain vinegar (which has yeast), soy, and other ingredients you may have eliminated during the 5-day quick start.

If you run into a situation where you haven't yet reintegrated certain ingredients called for in one of the daily recipes, simply substitute another recipe in Phase I. All of the recipes in Phase I can be eaten any day you are on the program, so it is perfectly safe for you to do this.

The key is to check for ingredients you haven't reintegrated yet and to make sure you don't accidentally eat them. That way you will be able to systematically test for foods to which you are sensitive.

Now it's time to congratulate yourself. You have completed the 5-day quick start and given yourself an even more powerful experience with health and weight loss than what is possible with the basic UltraMetabolism program. I hope that you can enjoy the benefits of additional weight loss and an enhanced feeling of vital health and that you use these as motivators for continuing with the rest of the UltraMetabolism Prescription.

Summary of the Quick Start and UltraMetabolism Programs

What follows is a simple reference guide for the quick start and UltraMetabolism programs. This will give you an easy way to make sure you are incorporating the two programs correctly to maximize your health and weight loss benefits.

- **Step 1:** Complete the preparation week. Follow the guidelines in chapter 19 to eliminate toxic substances in your diet and prepare for all the goodness to come.
- **Step 2:** Take the "Inflammation and Toxicity Quiz." On the evening before you begin the quick start (the last night of the preparation week), take the quiz again to assess how toxic and inflamed you are and learn how much the quick start will help you.
- **Step 3:** Begin the 5-day quick start. Follow the daily eating guidelines on pages 249–251 in place of the first five days of Phase I of the UltraMetabolism Prescription. Remember to start on a Sunday to make the program as easy as possible.
- **Step 4:** Complete the 5-day quick start and take the quiz again. Follow the quick-start program for five days. On the evening of the fifth day, take the quiz again to see how much your health has improved in five short days. Get ready to start reintegrating foods and move into Phase I tomorrow.
- **Step 5:** Proceed to Phase I and begin reintegrating foods. Using the guidelines on page 353 of this appendix, begin reintegrating

the foods you eliminated as you move into Phase I. Follow the daily dietary suggestions in chapter 20 and complete the remaining two days and two weeks of Phase I. Remember to reintegrate foods slowly. Also remember to watch for ingredients in the Phase I recipes that you may not have reintegrated yet. If you run across one, simply substitute another recipe.

✤ **Step 6:** Proceed to Phase II. Once you have completed Phase I, follow the guidelines in chapter 21 to move into Phase II.

✤ **Step 7:** Experience an UltraMetabolism for life! Phase II never really ends. You can continue eating this way for the rest of your life. Follow the UltraMetabolism lifestyle and experience Ultra-Wellness for the rest of your life.

Herbs and Supplements

Herbal Remedies

A Note on Herbs

The quality and beneficial effects of an herb depends on the field in which it was grown, the harvesting, shipping, and storage, the processing of the raw materials, the variation in active ingredients from batch to batch, and unintentional contaminants, as well as the form and dose in which it is ingested. Given those variables, it is hard to imagine you can find an herb of good quality. It often takes homework, but I tend to rely on these companies, which do most of the homework for me through their fastidious standards. Please see www.ultrametabolism.com for more information on optimal sources for herbs.

Ashwagandha

This Indian herb is a commonly used stress reducer and immune booster or adaptogen. The dose is 150mg once or twice a day.

Capsaicin

This is an extract of cayenne pepper and can be used to reduce cholesterol and lower blood pressure, but is often used as a topical painkiller for arthritis. There is actually a prescription medication for topical application. Capsules of cayenne, or preparations containing cayenne, can be a useful adjunct to treating inflammation. The supplement dose is measured in heat units, often up to 100,000 a day.

Cinnamon

Recent studies have found cinnamon to be a potent aid in normalizing blood sugar in people with diabetes at doses of 1 to 2 grams. It can be taken as a supplement in capsule form. Getting it on French toast is not going to work!

Cocoa

Emerging as the king of antioxidant and anti-inflammatory polyphenols, not to mention PEA or phenylethylamine that mimics the love molecules in your brain, and the source of OEA, the special fat that turns on your metabolism, a little dark chocolate is good for the body and the soul.

Enzymes (bromelain and other proteolytic enzymes)

Bromelain is the most well known anti-inflammatory enzyme. It is found in pineapple stems and helps in many inflammatory problems including trauma and muscle injury as well as asthma, arthritis, and colitis. The dose is about 600 mg a day taken in divided doses between meals. Take a capsule rated at least 2000 GDU.

Fenugreek

Fenugreek seeds, once crushed into a powder and taken in relatively large amounts (25 to 50 grams a day), can significantly lower blood sugar and blood fats. Smaller amounts can also be useful. For people who have significant sugar problems, this can be a helpful herb.

Ginger

This is another great addition to your daily meals (but not necessarily in gingerbread cookies, ginger ale, or pumpkin pie!). It can help thin the blood, lower cholesterol, and prevent nausea as well as act as a potent anti-inflammatory. Use fresh ginger in cooking or take ginger two to four 500mg capsules per day. Look also for extracts standardized to 5 percent gingerols. Or just use it in everyday cooking.

Ginseng

Panax ginseng—Chinese or
  Korean
Panax quinquefolius—
  American ginseng

There is some evidence that ginseng may help regulate insulin and blood sugar, and enhance immunity and adrenal function, as well as improve your ability to cope with stress. Take 200mg of a standardized extract twice daily.

Green tea

This everyday beverage in China contains a class of compounds called polyphenols, a type of flavonoid including epigallactocatechins that boost liver detoxification and reduce cholesterol, inflammation, and oxidative stress, and can help prevent cancer and heart disease. It is also thermogenic and may help increase metabolism and help with weight loss.

The dose is about 240 to 320mg of the phenols per day. Look for brands containing green tea extract standardized to 80 percent total polyphenols and 55 percent epigallocatechin gallate.

Licorice

The Chinese, Greeks, and Native Americans have used this for centuries as a medicinal plant. It has potent anti-inflammatory effects that come from its main chemical component, glycyrrhizin; it helps to balance the adrenal glands and has antiviral properties. This does not include the junk-food type of licorice. Use three to six capsules a day, but be careful if you have high blood pressure and fluid retention. You must monitor for these side effects, but for those with adrenal exhaustion manifested by low blood pressure and dizziness, this can be just the trick. The dose is 900mg a day.

Milk thistle (silymarin)

Milk thistle is also known as silymarin. This is an old herbal remedy for liver disease that has been shown in controlled studies to improve liver function in people with alcoholic and infectious hepatitis. It works by increasing the synthesis of glutathione as an antioxidant and by increasing the rate of liver tissue regeneration. The standard dose is 70 to 210mg a day. I recommend the standardized form from Germany that has been proven in research trials.

Quercitin (fruit and vegetable rind)

Quercetin is the king of bioflavonoids. This potent plant bioflavonoid from onions and garlic and fruit has anti-inflammatory and antihistamine properties. In part it acts by preventing the release of histamine from mast cells (special white blood cells that contain histamine) and can help with food and environmental allergies. For food allergy take 500mg about fifteen minutes before meals to reduce food reactions.

Rhodiola

This relative newcomer to the herbal world of adaptogens is called Arctic root. It has few side effects and gently boosts energy and increases your resistance to stress. The dose is 100 mg to 200 mg twice a day standardized to 3 percent rosavin.

Siberian ginseng

This Arctic tonic was used by Russian cosmonauts to boost their mental and immune functioning under the stress of space travel. It is less stimulating than Chinese or American ginseng, and can be a very good immune and adrenal tonic for regular use. The dose is about 200mg of a standardized extract two or three times a day.

Turmeric

This is the yellow spice commonly found in curry or yellow rice. It can be a powerful ally against inflammation and oxidative stress, useful in many inflammatory conditions. The dose is about 400 to 1200mg a day of curcumin, which is the active ingredient.

Adding Supplements

Note on supplements:

There are many special supplements that have multiple benefits across multiple keys. That is because the body runs on some very specific raw materials and when things get rough (as in illness or obesity) more of these raw materials are needed to keep up with the demand. Rather than repeating the supplements over and over in each chapter, I highlight them in the chapter where they have the most benefit. Just remember that some special supplements can help reduce inflammation and oxidative stress, improve mitochondrial function, and help you detoxify.

Acetyl L-carnitine

This is another important amino acid that helps transport fat into the mitochondria for burning. People who have genetic problems with their mitochondria—such as diabetics or those with insulin resistance—can be helped by this nutrient. It has been shown to help prevent damage and improve the activity of mitochondria in aging.

- ✣ 500 to 2000mg a day

Alpha lipoic acid

Alpha lipoic acid is a powerful antioxidant and metabolic booster that has been shown to reduce blood sugar and prevent diabetic complications. It helps recycle the antioxidants, including vitamin C, E, and beta-carotene, making them more available to quench free radicals. It is also important as an antioxidant defense for the mitochondria and helps improve energy metabolism.

- ✣ 100 to 300mg twice a day to help improve insulin resistance

Amino acids: the building blocks of protein

Some amino acids are important parts of the energy production cycle in the mitochondria. They also help detoxification. Providing enough of these is important. Occasionally supplements can help.

Take a few grams of a balanced amino acid powder daily or take:

- ✣ **taurine** 500mg twice a day and glycine 500mg twice a day to help with detoxification.
- ✣ **arginine,** an essential amino acid that helps to dilate arteries, improves blood flow, and lowers blood pressure, and may improve insulin resistance. The dose is 500 to 2000mg a day.
- ✣ **aspartic acid combined with magnesium** 200 to 400mg a day to help boost energy production in the mitochondria.

B complex vitamins

Our need for B vitamins increases with stress. They help us improve the metabolism of stress hormones so we can get rid of them. They are part of your basic multivitamin, but taking an additional B complex in times of increased stress can be helpful.

Bioflavonoids (citrus, pine bark, grape seed, green tea)

These compounds are the key plant compounds or pigments—about 4000 in total—that provide the color for our plants. They are often combined with vitamin C in supplements to prevent their destruction or oxidation. Familiar compounds rich in bioflavonoids include citrus foods, gingko biloba, bilberry, genistein from soy, red wine (resveratrol), and green tea (catechins). They can all be used as part of a diet or in supplements to reduce inflammation and oxidative stress.

Here are some of the more important ones:

- **Quercetin**—the king of bioflavonoids. This potent plant bioflavonoid from onions and garlic has anti-inflammatory and antihistamine properties. In part it acts by preventing the release of histamine from mast cells (special white blood cells that contain histamine) and can help with food and environmental allergies. For food allergy take 500mg about fifteen minutes before meals to reduce food reactions.
- **Pycnogenol or grape seed extract**—these contain potent anti-inflammatory and antioxidant bioflavonoids called proanthocyanidins. The dose is between 50 and 300mg a day.
- **Rutin**—this is a powerful bioflavonoid helpful for

inflammation in the blood vessels.

Coenzyme Q10

CoQ10 is a part of a critical step in the mitochondria involved in energy production, and helps boost metabolism. It also acts as an antioxidant. People who take the class of cholesterol-lowering drugs called statins deplete their CoQ10 levels because the same step in the body that produces cholesterol also produces CoQ10. In Parkinson's disease, the mitochondria are damaged by toxins. CoQ10 has been shown to slow or stop the progression of Parkinson's when given in high doses (1200mg a day), without any side effects.

- 50 to 1200mg a day

Creatine Powder

Creatine is another amino acid or building block of protein used for energy production in the mitochondria. It is commonly used by body builders to increase muscle mass. Recently, research has shown it effective in preserving or increasing muscle mass in muscle-wasting diseases such as muscular dystrophy and ALS (amyotrophic lateral sclerosis, or Lou Gehrig's disease). It can be used to build muscle and improve stamina and help your mitochondria produce energy.

- 2 to 4 grams a day

D-Ribose

This sugar is the raw material for energy production and the creation of ATP in the cells. It helps your mitochondria generate more energy along with carnitine and CoQ10 and magnesium. It comes in a powder.

- ❖ The dose is 5 grams mixed with water once or twice a day.

Extra-buffered vitamin C with mineral ascorbates

We have no ability to increase our vitamin C production like most mammals can under stress. Taking extra vitamin C helps support your adrenal glands and your immune system during stress. During any state of toxicity, vitamin C needs are increased. There has also been good evidence that those with higher vitamin C levels excrete more heavy metals such as lead and mercury.

- ❖ 1000 to 4000mg a day in powder, capsule, or tablets during periods of increased detoxification. This can cause loose stools. If it does, just reduce the dose or stop it.

GLA or gamma linolenic acid (evening primrose oil)

This is one of the good and essential omega-6 fats that our body can't produce. It is helpful in reducing inflammation and can help lower blood pressure and cholesterol, as well as improve fat metabolism in people with diabetes. Taking one or two grams of evening primrose oil is a good way to get these essential fatty acids.

- ❖ One to two grams of evening primrose oil twice a day can reduce inflammation and improve metabolism.
- ❖ Take 500mg to 1000mg twice a day.

Magnesium

Magnesium is the ultimate relaxation mineral. Under times of stress more magnesium is excreted in our urine. If we are deficient in magnesium we tighten up everywhere—getting headaches, constipation, palpitations, muscle cramps, and irritability. Additional magnesium is important for many people.

- ❖ If you tend to be constipated, use magnesium citrate 150mg once to twice a day; if you tend to have a sensitive stomach and loose bowels, use magnesium glycinate 150mg once to twice a day. If you take too much and have loose bowels, just reduce the dose.

N-acetylcysteine (NAC)

This amino acid–derived, sulphur-containing molecule is a key part of the way the body manufactures glutathione. In fact, it is used in emergency rooms to treat Tylenol overdose and liver failure, and to prevent kidney failure for patients in hospitals getting X-rays or angiograms using dye, which can damage the kidneys. Taking this as a supplement can boost the body's own glutathione, which is one of the critical antioxidants that protect the mitochondria.

❖ 500 to 2000mg a day

NADH

This is another little molecule the body makes that can get depleted, part of the critical energy production process in the mitochondria. This has been used effectively in patients with Chronic Fatigue Syndrome—a condition of malfunctioning and poisoned mitochondria. It can have an energy-boosting and alertness effect like caffeine without the jitters.

❖ 5 to 20mg a day

PGX (PolyGlycopleX) or konjac root fiber

This is the special superfiber that is very viscous and soaks up fat, sugar, and water in the gut and reduces the overall glycemic load of any meal you eat. I have found this a uniquely powerful, safe way to promote weight loss and lower blood sugar and cholesterol in many patients.

❖ Take as 2 to 4 capsules before every meal.

Probiotics

These are found in fermented foods like yogurt, but to obtain optimal amounts use probiotic supplements, which include lactobacillus acidophilus, lactobacillus rhamnosis, and bifidobacteria. By restoring the normal gut flora, they reduce overall immune activation, and have been proven effective in many inflammatory diseases including asthma, eczema, rhinitis, and inflammatory bowel disease. They work by balancing the GALT or gut associated lymphoid tissue.

These are found in the refrigerated section of most health food stores.

❖ Look for capsules containing 5 to 10 billion live organisms. Higher doses are often used in severe inflammatory conditions.

Zinc

Zinc is important for almost every function of the body, as well as in the normal function of the adrenal glands, modulating stress hormones and supporting the immune system. It is also one of the most common nutritional deficiencies worldwide. Your multivitamin will support your basic needs, but additional stress increases your needs.

❖ Take an extra 15 to 30mg of zinc a day. Ideal forms are zinc citrate, picolinate, aspartate, or chelate.

Activating the Relaxation Response

(See www.ultrametabolism.com for more resources)

There are many wonderful resources available to help you activate the relaxation response and reduce stress. Below is a selection of some of the best sources of CDs and lifestyle products such as biofeedback tools and saunas.

Health Journeys
www.healthjourneys.com
Resources for self-healing including guided imagery tapes

Natural Journeys
www.naturaljourneys.com
Healthy lifestyle DVDs and videos including Pilates, yoga, tai chi, fitness, meditation, and self-healing

The Relaxation Company
www.therelaxationcompany.com
Music and relaxation CDs

Mental Yoga
www.mentalyoga.com
A new cognitive process that teaches you to understand and transform the thought patterns that create stress (and lead to disease and weight gain). This can be learned online.

Tapes and Resources for Yoga, Breathing, and Meditation

Padma Media: Products That Support An Awakened Life
www.susanpiver.com/wordpress/padma-media.com

Tools for Healthy Living and Relaxation

Resperate
www.resperate.com
A personal small biofeedback device to train yourself to relax

Sunlight Saunas—Source of Far Infrared Saunas
www.sunlightsaunas.com

Pangea Organics
www.pangeaorganics.com
Organic personal care products

ACKNOWLEDGMENTS

Writing a book is like raising a child. Like a child, it brings tremendous joys and the deepest struggles. And like a child, hundreds of people are involved in raising it in ways both large and small.

It took a community to write this book. That community includes all the scientists who worked thanklessly to understand the mysteries of the human body and all my patients who trusted me, then worked with me to find the answers to their health and weight struggles. They taught me more than they realized.

My agent, Richard Pine, with patience, clarity, insight, and uncommon directness, has guided this book from the beginning. Beth Wareham, my editor, and all my friends and supporters at Simon and Schuster have supported me from the beginning to the end. Marc Stockman tirelessly and enthusiastically helped me harness the power of this information to make it more useful for everyone. And Spencer Smith helped me shape it with great humor and patience. Sonya Irish traced my thought lines through hundreds of research papers. And my mother, Ruth, read every word to make sure it was clear to everyone what I meant.

The community that helped me extends to well over a hundred people, all of whom I unfortunately can't name here. You know who you are—thank you, thank you, and thank you. I must mention a few special people who have inspired, helped, and supported me: Jeff Bland, Sidney Baker, my friends at the Institute for Functional Medicine, Mel and Enid Zuckerman and Jerry Cohen, Richard Butler, David Ludwig, Marc David, David Piver, Jonathan Kalman, David Eisenberg, Peter Libby, and the list goes on.

Thank you Kadan Swift, Andrea Bushey, Marie Breen, and Katie Breen for testing and tasting all the recipes. Laurie Erickson provided invaluable enthusiasm and assistance with ideas for the recipes.

My life would have taken a different turn if not for Kathie Swift, my nutritionist and coconspirator, who first inspired me to ask the questions that led to this book. Her tireless hours of assistance and work, plus her dedication to our patients, to food, and to this book, in more ways than I can count, were essential to its creation. Thank you.

Last, and most important, my family has put up with the dangers of my passion (early mornings, late nights, and too many absences to remember). I could not have done this without all your love and belief in what I am doing. Thank you, Pier, Rachel, Misha, Thor, Ace, and Max.

NOTES

Epigraph

1. Bennett WI. Beyond overeating. *NEJM*. 1995;332:673–674.
2. Bjorntorp PA. Overweight is risking fate. *Baillieres Best Pract Res Clin Endocrinol Metab*. 1999 Apr;13(1):47–69. Review.

Introduction

1. Eckel R. The dietary approach to obesity: Is it the diet or the disorder? *JAMA*. 2005;293(1):96–97.
2. Hyman M. Paradigm shift: The end of "normal science" in medicine. Understanding function in nutrition, health, and disease. *Altern Ther Health Med*. 2004 Sep–Oct;10(5).
3. Mokdad AH. Actual causes of death in the United States, 2000. *JAMA* 2004;291:1238–1245.
4. Fontaine KR, Redden DT, Wang C, Westfall AO, Allison DB. Years of life lost due to obesity. *JAMA*. 2003 Jan 8;289(2):187–193.
5. Sarlio-Lahteenkorva S, Rissanen A, Kaprio J. A descriptive study of weight loss maintenance: 6 and 15 year follow-up of initially overweight adults. *Int J Obes Relat Metab Disord*. 2000 Jan;24(1):116–22.
6. Dansinger ML, Gleason JA, Griffith JL, Selker HP, Schaefer EJ. Comparison of the Atkins, Ornish, Weight Watchers, and Zone diets for weight loss and heart disease risk reduction: a randomized trial. *JAMA*. 2005 Jan 5;293(1):43–53.
7. Eckel RH. The dietary approach to obesity: Is it the diet or the disorder? *JAMA*. 2005 Jan 5;293(1):96–97.
8. Planck M. *Scientific Autobiography and Other Papers,* trans. F. Gaynor (New York: Philosophical Library, 1949), pp. 33–34.
9. Satyanarayan K, Shoskes DA. A molecular injury-response model for the understanding of chronic disease. *Molecular Medicine Today*. 1997 Aug;3(8):331–34. Review.
10. Frist WH. Shattuck Lecture: Health care in the 21st century. *NEJM*. 2005 Jan 20;352(3).

Chapter 2: The Calorie Myth: All Calories Are Created Equal

1. Ludwig D. Dietary glycemic index and obesity. *J Nutr*. 2000;130:280S–283S.
2. Greene P. Pilot 12-week feeding weight loss comparison: Low fat vs. low carbohydrate diets. Abstract 95. Presented at the North American Association for the Study of Obesity's 2003 Annual Meeting.

3. Ludwig D. High glycemic index foods, overeating and obesity. *Pediatrics.* 1999;102(3):26.
4. Kaput J, Rodriguez RL. Nutritional genomics: the next frontier in the post-genomic era. *Physiol Genomics.* 2004 Jan 15;16(2):166–177. Review.

Chapter 3: The Fat Myth: Eating Fat Makes You Fat

1. Taubes G. The soft science of dietary fat. *Science,* 2001 Mar 30;291:2536–2545.
2. Willett W. Dietary fat is not a major determinant of body fat. *Am J Med.* 2002;113(9B);47S–59S.
3. Foster GD. A randomized trial of a low carbohydrate diet for obesity. *NEJM.* 2003;348:2082–2090. Samaha FF. A low-carbohydrate as compared with a low-fat diet in severe obesity. *NEJM.* 2003;348:2074–2081.
4. Olshansky SJ, Passaro DJ. A potential decline in life expectancy in the United States in the 21st century. *NEJM.* 2005:352(11):1138–1145.
5. De Lorgeil M. Mediterranean alpha-linolenic acid rich diet in secondary prevention of coronary heart disease. *Lancet.* 1994;343:1454–1459.
6. Knoops KT, de Groot LC, Kromhout D, Perrin AE, Moreiras-Varela O, Menotti A, van Staveren WA. Mediterranean diet, lifestyle factors, and 10-year mortality in elderly European men and women: the HALE project. *JAMA.* 2004 Sep 22;292(12):1433–1439.
7. Mensink RP, Zock PL, Kester A, Katan MB. Effects of dietary fatty acids and carbohydrates on the ratio of serum total to HDL cholesterol and on serum lipids and apolipoproteins: A meta-analysis of 60 controlled trials. *Am J Clin Nutr* 2003;77:1146–1155.
8. Evans RM, Barish GD, Wang YX. PPARs and the complex journey to obesity. *Nat Med.* 2004 Apr;10(4):355–361. Review.
9. Bray GA, Lovejoy JC, Smith SR, DeLany JP, Lefevre M, Hwang D, Ryan DH, York DA. The influence of different fats and fatty acids on obesity, insulin resistance and inflammation. *J Nutr.* 2002;132(9):2488–2491.
10. Chambrier C, Bastard JP, Rieusset J, Chevillotte E, Bonnefont-Rousselot D, Therond P, Hainque B, Riou JP, Laville M, Vidal H. Eicosapentaenoic acid induces mRNA expression of peroxisome proliferator-activated receptor gamma. *Obes Res.* 2002;10(6):518–525.
11. Kang K, Liu W, Albright KJ, Park Y, Pariza MW. Trans-10, cis-12 CLA inhibits differentiation of 3T3-L1 adipocytes and decreases PPAR gamma expression. *Biochem Biophys Res Commun.* 2003;303(3):795–799.

Chapter 4: The Carb Myth: Eating Low Carb or No Carb Will Make You Thin

1. Bagnulo, J. Cutting through the Carbohydrate Confusion. *Alternative Therapies in Health and Medicine.* Sept-Oct 2005;11(5):18–20.
2. Boyce VL, Swinburn BA. The traditional Pima Indian diet: Composition and adaptation for use in a dietary intervention study. *Diabetes Care.* 1993 Jan;16(1):369–371.

3. Olshansky SJ, Passaro DJ, Hershow RC, Layden J, Carnes BA, Brody J, Hayflick L, Butler RN, Allison DB, Ludwig DS. A potential decline in life expectancy in the United States in the 21st century. *NEJM.* 2005 Mar 17;352(11):1138–1145.
4. Ludwig DS, Pereira MA, Kroenke CH, Hilner JE, Van Horn L, Slattery ML, Jacobs DR Jr. Dietary fiber, weight gain, and cardiovascular disease risk factors in young adults. *JAMA.* 1999 Oct 27;282(16):1539–1546.

Chapter 5: The Sumo Wrestler Myth: Skipping Meals Helps You Lose Weight

1. Wyatt HR. Long-term weight loss and breakfast in subjects in the National Weight Control Registry. *Obesity Res.* 2002 Feb;10(2):78–82.
2. Farshchi HR, Taylor MA, Macdonald IA. Deleterious effects of omitting breakfast on insulin sensitivity and fasting lipid profiles in healthy lean women. *Am J Clin Nutr.* 2005 Feb;81(2):388–396.

Chapter 6: The French Paradox Myth: The French Are Thin Because They Drink Wine and Eat Butter

1. Nestle M. Food politics: How the food industry influences nutrition and health. University of California Press, 2002.
2. Rozin P. The ecology of eating: Smaller portion sizes in France than in the United States help explain the French paradox. *Psychol Sci.* 2003 Sep;14(5): 450–457.
3. Samara JN. Patterns and trends in food portion sizes, 1977–1998. *JAMA.* 2003:289:450–453.
4. Mokdad AH, Marks JS, Stroup DF, Gerberding JL. Actual causes of death in the United States, 2000. *JAMA* 2004 Mar 10;291(10):1238–1245. Review.
5. Gershon M. The second brain: A groundbreaking new understanding of nervous disorders of the stomach and intestine, *Perennial.* 1999.
6. David M. *The Slow Down Diet: Eating for Pleasure, Energy and Weight Loss.* Healing Arts Press, 2005.

Chapter 7: The Protector Myth: Government Policies and Food Industry Regulations Protect Our Health

1. Nestle M. The Ironic Politics of Obesity. *Science.* 2003 Feb 7;299.
2. Gallo AE. Food advertising in the United States. In Frazao E, ed., *America's Eating Habits: Changes & Consequences.* Washington, D.C.: USDA, 1999, pp. 173–180.
3. Egan T. Failing farmers learn to profit from federal aid. *The New York Times,* December 24, 2000: A1, A20.
4. Mokdad AH. Actual causes of death in the United States, 2000. *JAMA.* 2004;291:1238–1245.
5. Cordain L, Eaton SB, Sebastian A, Mann N, Lindeberg S, Watkins BA, O'Keefe

JH, Brand-Miller J. Origins and evolution of the Western diet: Health implications for the 21st century. *Am J Clin Nutr.* 2005 Feb;81(2):341–354. Review.

Chapter 9: Control Your Appetite: Harnessing Your Brain Chemistry for Weight Loss

1. Studer M, Briel M, Leimenstoll B, Glass TR, Bucher HC. Effect of different antilipidemic agents and diets on mortality: A systematic review. *Arch Intern Med.* 2005 Apr 11;165(7):725–730. Review.

2. Pereira MA, Swain J, Goldfine AB, Rifai N, Ludwig DS. Effects of a low-glycemic load diet on resting energy expenditure and heart disease risk factors during weight loss. *JAMA.* 2004 Nov 24;292(20):2482–2490.

3. Vuksan V, Sievenpiper JL, Owen R, Swilley JA, Spadafora P, Jenkins DJ, Vidgen E, Brighenti F, Josse RG, Leiter LA, Xu Z, Novokmet R. Beneficial effects of viscous dietary fiber from Konjac-mannan in subjects with the insulin resistance syndrome: results of a controlled metabolic trial. *Diabetes Care.* 2000 Jan;23(1):9–14.

4. Juntunen KS, Laaksonen DE, Poutanen KS, Niskanen LK, Mykkänen HM. High-fiber rye bread and insulin secretion and sensitivity in healthy postmenopausal women. *Am J Clin Nutr.* 2003 Feb;77:385–391.

5. Ludwig DS. The glycemic index: physiological mechanisms relating to obesity, diabetes, and cardiovascular disease. *JAMA.* 2002 May 8;287(18):2414.

6. Bray GA, Nielsen SJ, Popkin BM. Consumption of high-fructose corn syrup in beverages may play a role in the epidemic of obesity. *Am J Clin Nutr.* 2004 Apr;79(4):537–543. Review.

7. Brownell KD, Horgen KB. *Food fight: The inside story of America's obesity crisis— and what we can do about it.* New York: McGraw-Hill, 2003.

8. Smith JD, Terpening CM, Schmidt SO, Gums JG. Relief of fibromyalgia symptoms following discontinuation of dietary excitotoxins. *Ann Pharmacother.* 2001 Jun;35(6):702–706.

9. Lavin JH et al. The effect of sucrose- and aspartame-sweetened drinks on energy intake, hunger and food choice of female, moderately restrained eaters. *Int J Obes.* 1997;21:37–42.

10. Tordoff MG, Alleva AM. Oral stimulation with aspartame increases hunger. *Physiol Behav.* 1990;47:555–559.

11. Sharma RP, Coulombe Restio A Jr. Effects of repeated doses of aspartame on serotonin and its metabolite in various regions of the mouse brain. *Food Chem Toxicol.* 1987;25(8):565–568.

12. Camfield PR, et al. Aspartame exacerbates EEG spike-wave discharge in children with generalized absence epilepsy: A double-blind controlled study. *Neurology.* 1992;42:1000–1003.

13. Walton RG, et al. Adverse reactions to aspartame: Double-blind challenge in patients from a vulnerable population. *Biol Psychiatry.* 1993;34(1–2):13–17.

14. Van Den Eeden SK, et al. Aspartame ingestion and headaches: A randomized, crossover trial. *Neurology.* 1994;44:1787–1793.

15. Lipton RB, et al. Aspartame as a dietary trigger of headache. *Headache.* 1989; 29(2):90–92; http://www.dorway.com/peerrev.html.

16. Farshchi HR, Taylor MA, Macdonald IA. Beneficial metabolic effects of regular meal frequency on dietary thermogenesis, insulin sensitivity, and fasting lipid profiles in healthy obese women. *Am J Clin Nutr.* 2005 Jan;81(1):16–24.

17. Jenkins DJ, Wolever TM, Vuksan V, Brighenti F, Cunnane SC, Rao AV, Jenkins AL, Buckley G, Patten R, Singer W, et al. Nibbling versus gorging: Metabolic advantages of increased meal frequency. *NEJM.* 1989;321(14):929–934.

18. Farshchi HR, Taylor MA, Macdonald IA. Deleterious effects of omitting breakfast on insulin sensitivity and fasting lipid profiles in healthy lean women. *Am J Clin Nutr.* 2005 Feb;81(2):388–396.

19. De Castro JM. The time of day of food intake influences overall intake in humans. *J Nutr.* 2004 Jan;134(1):104–111.

Chapter 10: Subdue Stress: How Stress Makes You Fat and Relaxing Makes You Thin

1. Chaouloff F, Laude D, Merino D, Serruier D, Elghozi JL. Peripheral and central consequences of immobilization stress in genetically obese Zucker rats. *Am J Physiol.* 1989 Feb;256(2Pt2):R435–442.

2. Tull ES, Sheu YT, Butler C, Cornelious K. Relationships between perceived stress, coping behavior and cortisol secretion in women with high and low levels of internalized racism. *J Natl Med Assoc.* 2005 Feb;97(2):206–212.

3. Landen M, Baghaei F, Rosmond R, Holm G, Bjorntorp P, Eriksson E. Dyslipidemia and high waist-hip ratio in women with self-reported social anxiety. *Psychoneuroendocrinology.* 2004 Sep;29(8):1037–1046.

4. Rosmond R, Dallman MF, Bjorntorp P. Stress-related cortisol secretion in men: relationships with abdominal obesity and endocrine, metabolic and hemodynamic abnormalities. *J Clin Endocrinol Metab.* 1998 Jun;83(6):1853–1859.

5. Marniemi J, Kronholm E, Aunola S, Toikka T, Mattlar CE, Koskenvuo M, Ronnemaa T. Visceral fat and psychosocial stress in identical twins discordant for obesity. *J Intern Med.* 2002 Jan;251(1):35–43.

6. Pawlow LA, O'Neil PM, Malcolm RJ. Night eating syndrome: Effects of brief relaxation training on stress, mood, hunger, and eating patterns. *Int J Obes Relat Metab Disord.* 2003 Aug;27(8):970–978.

7. Vicennati V, Ceroni L, Gagliardi L, Gambineri A, Pasquali R. Comment: Response of the hypothalamic-pituitary-adrenocortical axis to high-protein/fat and high-carbohydrate meals in women with different obesity phenotypes. *J Clin Endocrinol Metab.* 2002 Aug;87(8):3984–3988.

8. Kristal A, Littman A, Benitex D, White E. Yoga practice is associated with an attenuated weight gain in healthy, middle-aged men and women. *Altern Ther Health Med.* 2005;11(4):28–33.

9. Nguyen Y, Naseer N, Frishman WH. Sauna as a therapeutic option for cardiovascular disease. *Cardiol Rev.* 2004 Nov–Dec;12(6):321–324.

Chapter 11: Cool the Fire of Inflammation: Hidden Fires That Make You Fat

1. Visser M. Elevated C-reactive protein levels in overweight and obese adults. *JAMA*. 1999 Dec 8;282(22):2131–2135.
2. Mozaffarian D, Pischon T, Hankinson SE, Rifai N, Joshipura K, Willett WC, Rimm EB. Dietary intake of trans fatty acids and systematic inflammation in women. *Am J Clin Nutr*. 2004 Apr;79(4):606–612.
3. Pradhan AD. C-reactive protein, interleukin 6, and risk of developing type 2 diabetes mellitus. *JAMA*. 2001 Jul 18;286(3):327–334.
4. Liu S, Manson JE, Buring JE, Stampfer MJ, Willett WC, Ridker PM. Relation between a diet with a high glycemic load and plasma concentrations of high-sensitivity C-reactive protein in middle-aged women. *Am J Clin Nutr*. 2002 Mar;75(3):492–498.
5. Evans RM, Barish GD, Wang YX. PPARs and the complex journey to obesity. *Nat Med*. 2004 Apr;10(4):355–361. Review.
6. Esposito K, Pontillo A, Di Palo C, Giugliano G, Masella M, Marfella R, Giugliano D. Effect of weight loss and lifestyle changes on vascular inflammatory markers in obese women: A randomized trial. *JAMA*. 2003 Apr 9;289(14): 1799–1804.
7. Ajani UA, Ford ES, Mokdad AH. Dietary fiber and C-reactive protein: findings from national health and nutrition examination survey data. *J Nutr*. 2004 May;134(5):1181–1185.
8. Jenkins DJ. Effects of a dietary portfolio of cholesterol-lowering foods vs. lovastatin on serum lipids and C-reactive protein. *JAMA*. 2003 Jul 23;290(4): 502–510.
9. Smith JK. Long-term exercise and atherogenic activity of blood mononuclear cells in persons at risk of developing ischemic heart disease. *JAMA*. 1999 May 12;281(18):1722–1727.
10. Church TS. Reduction of C-reactive protein levels through use of a multivitamin. *Am J Med*. 2003 Dec 15;115(9):702–707.
11. Chambrier C, Bastard JP, Rieusset J, Chevillotte E, Bonnefont-Rousselot D, Therond P, Hainque B, Riou JP, Laville M, Vidal H. Eicosapentaenoic acid induces mRNA expression of peroxisome proliferator-activated receptor gamma. *Obes Res*. 2002 Jun;10(6):518–525.
12. LoVerme J, Fu J, Astarita G, La Rana G, Russo R, Calignano A, Piomelli D. The nuclear receptor peroxisome proliferator-activated receptor-alpha mediates the anti-inflammatory actions of palmitoylethanolamide. *Mol Pharmacol*. 2005 Jan;67(1):15–19. Epub 2004 Oct 1.
13. Sies H. Cocoa polyphenols and inflammatory mediators. *Am J Clin Nutr*. 2005 Jan;81(1 Suppl.):304S–312S. Review.
14. See Resources for the best places to buy the "weight loss" chocolate.
15. Isolauri E, Rautava S, Kalliomaki M. Food allergy in irritable bowel syndrome: new facts and old fallacies. *Gut*. 2004;53(10):1391–1393. Atkinson W, Sheldon TA, Shaath N, Whorwell PJ. Food elimination based on IgG antibodies in

irritable bowel syndrome: A randomised controlled trial. *Gut.* 2004;53 (10):1459–1464.

Chapter 12: Prevent Oxidative Stress or "Rust": Keep the Free Radicals from Taking Over

1. Evans JL, Goldfine IO, Maddux BA, Grodsky GM. Oxidative stress and stress-activated signaling pathways: A unifying hypothesis of type 2 diabetes. *Endocrine Reviews.* 23(5):599–622.
2. Engelhart, et al. Dietary intake of antioxidants and risk of Alzheimer disease. *JAMA.* 2002 Jun 26;287(24):3223–3229.
3. Sies H, Schewe T, Heiss C, Kelm M. Cocoa polyphenols and inflammatory mediators. *Am J Clin Nutr.* 2005 Jan;81(1):304S–312S.
4. Scalbert A, Johnson IT, Saltmarsh M. Polyphenols: Antioxidants and beyond. *Am J Clin Nutr.* 2005 Jan;81(1):215S–217S.

Chapter 13. Turn Calories into Energy: Increasing Your Metabolic Power

1. Heilbronn LK, Ravussin E. Calorie restriction and aging: Review of the literature and implications for studies in humans. *Am J Clin Nutr.* 2003 Sep;78: 361–369.
2. Petersen KF, Dufour S, Befroy D, Garcia R, Shulman GI. Impaired mitochondrial activity in the insulin-resistant offspring of patients with type 2 diabetes. *NEJM.* 2004;350:664–671.
3. Lowell BB, Shulman GI. Mitochondrial dysfunction and type 2 diabetes. *Science.* 2005 Jan 21;307(5708):384–387.
4. Levine JA, Lanningham-Foster LM, McCrady SK, Krizan AC, Olson LR, Kane PH, Jensen MD, Clark MM. Interindividual variation in posture allocation: Possible role in human obesity. *Science.* 2005 Jan 28;307(5709):584–586.
5. Tremblay A, Simoneau JA, Bouchard C. Impact of exercise intensity on body fatness and skeletal muscle metabolism. *Metabolism.* 1994 Jul;43(7):814–818.
6. Hagen TM, Liu J, Lykkesfeldt J, Wehr CM, Ingersoll RT, Vinarsky V, Bartholomew JC, Ames BN. Feeding acetyl-L-carnitine and lipoic acid to old rats significantly improves metabolic function while decreasing oxidative stress. *Proc Natl Acad Sci USA.* 2002 Feb 19;99(4):1870–1875. Erratum in: *Proc Natl Acad Sci USA* 2002 May 14;99(10):7184.

Chapter 14: Fortify Your Thyroid: Maximizing the Major Metabolism Hormone

1. Tuzcu A, Bahceci M, Gokalp D, Tuzun Y, Gunes K. Subclinical hypothyroidism may be associated with elevated high-sensitive C-reactive protein (low grade inflammation) and fasting hyperinsulinemia. *Endocr J.* 2005 Feb;52(1):89–94.
2. Persky VW, Turyk ME, Wang L, Freels S, Chatterton R Jr, Barnes S, Erdman J Jr,

Sepkovic DW, Bradlow HL, Potter S. Effect of soy protein on endogenous hormones in postmenopausal women. *Am J Clin Nutr.* 2002 Jan;75(1):145–153. Erratum in: *Am J Clin Nutr.* 2002 Sep;76(3):695.

3. Toscano V, Conti FG, Anastasi E, Mariani P, Tiberti C, Poggi M, Montuori M, Monti S, Laureti S, Cipolletta E, Gemme G, Caiola S, Di Mario U, Bonamico M. Importance of gluten in the induction of endocrine autoantibodies and organ dysfunction in adolescent celiac patients. *Am J Gastroenterol.* 2000 Jul;95(7):1742–1748.

4. Ellingsen DG, Efskind J. Effects of low mercury vapour exposure on the thyroid function in chloralkali workers. *J Appl Toxicol.* 2000 Nov–Dec;20(6):483–489.

5. Galletti PM, Joyet G. Effect of fluorine on thyroidal iodine metabolism in hyperthyroidism. *J Clin Endocrinol Metab.* 1958 Oct;18(10):1102–1110.

6. WJ, Pan Y; Johnson AR, et al. Reduction of chemical sensitivity by means of heat depuration, physical therapy and nutritional supplementation in a controlled environment. *J Nutr Env Med.* 1996;6:141–148.

7. Pelletier C, Imbeault P, Tremblay A. Energy balance and pollution by organochlorines and polychlorinated biphenyls. *Obes Rev.* 2003 Feb;4(1):17–24. Review.

8. Bland J. Nutritional Endocrinology, Normalizing Hypothalamus-Pituitary-Thyroid Axis Function, 2002 Seminar Series Syllabus.

9. Cooper DS. Subclinical Hypothyroidism. *NEJM.* 2001 Jul 26;345:260–265.

10. Gaby AR. Sub-laboratory hypothyroidism and the empirical use of Armour thyroid. *Altern Med Rev.* 2004 Jun;9(2):157–179.

11. Goglia F. Biological effects of 3,5-diiodothyronine (T(2)). *Biochemistry* (Moscow). 2005 Feb;70(2):164–172.

Chapter 15: Love Your Liver: Cleansing Yourself of Toxic Weight

1. Baillie-Hamilton PF. Chemical toxins: A hypothesis to explain the global obesity epidemic. *J Altern Complement Med.* 2002 Apr;8(2):185–192. Review.

2. U.S. Centers for Disease Control and Prevention. Second National Report on Human Exposure to Environmental Chemicals. www.cdc.gov/exposurereport/2nd/.

3. Pelletier C, Imbeault P, Tremblay A. Energy balance and pollution by organochlorines and polychlorinated biphenyls. *Obes. Rev.* 2003 Feb;4(1):17–24. Review.

4. Tremblay A, Pelletier C, Doucet E, Imbeault P. Thermogenesis and weight loss in obese individuals: A primary association with organochlorine pollution. *Int J Obes Relat Metab Disord.* 2004 Jul;28(7):936–939.

5. Imbeault P, Tremblay A, Simoneau JA, Joanisse DR. Weight loss–induced rise in plasma pollutant is associated with reduced skeletal muscle oxidative capacity. *Am J Physiol Endocrinol Metab.* 2002 Mar;282(3):E574–E579.

6. Duke University Integrated Toxicology Program, National Institute of Environmental Health Sciences/NIH/DHHS. Obesity: Developmental Origins

and Environmental Influences; www.niehs.nih.gov/multimedia/qt/dert/obesity/agenda.htm.

7. Beattie JH, Wood AM, Newman AM, Bremner I, Choo KH, Michalska AE, Duncan JS, Trayhurn P. Obesity and hyperleptinemia in metallothionein (-I and -II) null mice. *Proc Natl Acad Sci USA.* 1998 Jan 6;95(1):358–363.

8. Masuda A, Miyata M, Kihara T, Minagoe S, Tei C. Repeated sauna therapy reduces urinary 8-epi-prostaglandin F(2alpha). *Jpn Heart J.* 2004 Mar;45(2):297–303.

9. Biro S, Masuda A, Kihara T, Tei C. Clinical implications of thermal therapy in lifestyle-related diseases. *Exp Biol Med* (Maywood). 2003 Nov;228(10):1245–1249. Review.

INDEX

Mark Hyman, M.D., is editor in chief of *Alternative Therapies in Health and Medicine,* the most prestigious journal in the field of integrative medicine, and the medical editor of *Alternative Medicine Magazine: The Art and Science of Healthy Living.* Dr. Hyman is one of the leading experts in prevention, health promotion, and a new medical paradigm that addresses the causes and roots of our epidemic of chronic disease and promises to help transform our disease care system into a health care system.

Dr. Hyman's perspective was shaped by his experience with chronic illness: first a ruptured disc, back surgery, and chronic pain that was cured through acupuncture in China, and then chronic fatigue syndrome and mercury toxicity, which he cured by applying this new medical paradigm. This led him on an impassioned journey to seek new scientific and innovative models to address the health problems of the twenty-first century.

He combines the best of conventional and alternative medicine with a blend of science, integrity, intuition, and compassion in his medical practice in the Berkshire Mountains of Massachusetts. He is the coauthor of the *New York Times* bestseller recently published by Scribner, *UltraPrevention: The Six-Week Plan that Will Make You Healthy for Life; The Detox Box: A Program for Greater Health and Vitality,* recently published by Sounds True in 2004; *The 5 Forces of Wellness: The UltraPrevention System for Living an Active, Age-Defying, Disease-Free Life,* recently published by Nightingale; and *Nutrigenomics: The New Science of Health and Weight Loss,* an audio-learning series. His website www.drhyman.com empowers health care consumers by allowing them to take advantage of the medicine of the future today.

For nearly ten years he was co-medical director at Canyon Ranch Lenox, an internationally acclaimed health resort, which is a practice affiliated with Harvard University's Brigham and Women's Hospital. He graduated with a B.A. from Cornell University in 1982 and magna cum laude from the University of Ottawa School of Medicine in 1987, and he graduated from the University of San Francisco's program in Family Medicine at Community Hospital of Santa Rosa. He is board certified in family medicine.

He recently testified at the White House Commission on Complementary and Alternative Medicine, and met with and advised the surgeon general on a new diabetes prevention initiative. He is a contributor to the *Textbook of Functional Medicine* (Institute of Functional Medicine, 2006).

WHO ELSE WANTS TO JOIN
The UltraMetabolism Community
AND START THE SIMPLE PLAN FOR AUTOMATIC WEIGHT LOSS?

One of the wonderful things about the Internet is it allows me to stay in touch with many more people than I otherwise could. And at the same time, it also allows all of you—people who are going through the same struggles and successes while losing weight—to stay in touch with each other.

This is enormously important because, in my experience, I've found that if someone has a support system in place, their chances of losing weight are much, much higher than if they don't.

Therefore, I have put together an online community that will allow all of you to stay in touch with each other, providing much-needed support and encouragement to one another, and help us to get as many people on the UltraMetabolism plan as possible.

When you get your own free membership in the UltraMetabolism community, you'll be able to do all of the following:

- Share your best UltraMetabolism recipes with other members.
- Follow my blog as I post my latest updates to the Ultra-Metabolism plan.
- Track your weight-loss efforts on a weekly basis.
- Create your own food journal so you can track what you are eating.
- Post thoughts to your own blog on how your weight-loss efforts are going.
- Seek out and find other members so you create your own online buddies.
- Hold online chats with your buddies to swap advice on losing weight.
- Ask questions to other members via message boards.
- Send private messages to other members who are helping you to lose weight.

✢ Read success stories of others who have already lost weight on UltraMetabolism.

✢ Invite your friends, family, coworkers, etc., to join.

Writing this book is part of my contribution toward helping thousands of people lose weight and regain their health, but I need your help in touching millions—not just thousands—so we can collectively help our society turn the corner and win the fight against obesity.

To join, simply go to: http://www.ultrametabolism.com/join.

Also by Mark Hyman, M.D.

UltraPrevention (with Mark Liponis, M.D.)

The Detox Box

The 5 Forces of Wellness

Nutrigenomics: The New Science of Health and Weight Loss